Evolutionary Ethics and Contemporary Biology

How can the discoveries made in the biological sciences play a role in a discussion on the foundation of ethics? This book responds to this question by examining how evolutionism can explain and justify the existence of ethical normativity and the emergence of particular moral systems. Written by a team of philosophers and scientists, the essays collected in this volume deal with the limits of evolutionary explanations, the justifications of ethics, and methodological issues concerning evolutionary accounts of ethics, among other topics. They offer deep insights into the origin and purpose of human moral capacities and of moral systems.

Giovanni Boniolo is Full Professor of Logic and Philosophy of Science at the Firc Institute for Molecular Oncology (IFOM) of Milano. He is the author of many articles in international journals and books, including *Metodo e rappresentazioni del mondo; Il limite e il ribelle. Etica, naturalismo, darwinismo;* and, with P. Vidali, *Filosofia della scienza* and *Strumenti per ragionare.*

Gabriele De Anna is Lecturer in Philosophy at the University of Udine. He is the author of *Realismo metafisico e rappresentazione mentale* and *Il pensiero filosofico e politico di Sebastiano Di Apollonia.*

T0292194

CAMBRIDGE STUDIES IN PHILOSOPHY AND BIOLOGY

General Editor
Michael Ruse   *Florida State University*

Advisory Board
Michael Donoghue   *Yale University*
Jean Gayon   *University of Paris*
Jonathan Hodge   *University of Leeds*
Jane Maienschein   *Arizona State University*
Jesús Mosterín   *Instituto de Filosofía (Spanish Research Council)*
Elliott Sober   *University of Wisconsin*

Published Titles

Alfred I. Tauber   *The Immune Self: Theory or Metaphor?*

Elliott Sober   *From a Biological Point of View*

Robert Brandon   *Concepts and Methods in Evolutionary Biology*

Peter Godfrey-Smith   *Complexity and the Function of Mind in Nature*

William A. Rottschaefer   *The Biology and Psychology of Moral Agency*

Sahotra Sarkar   *Genetics and Reductionism*

Jean Gayon   *Darwinism's Struggle for Survival*

Jane Maienschein and Michael Ruse (eds.)   *Biology and the Foundations of Ethics*

Jack Wilson   *Biological Individuality*

Richard Creath and Jane Maienschein (eds.)   *Biology and Epistemology*

Alexander Rosenberg   *Darwinism in Philosophy, Social Science and Policy*

Peter Beurton, Raphael Falk, and Hans-Jörg Rheinberger (eds.)
The Concept of the Gene in Development and Evolution

David L. Hull   *Science and Selection*

James G. Lennox   *Aristotle's Philosophy of Biology*

Marc Ereshefsky   *The Poverty of the Linnaean Hierarchy*

Kim Sterelny   *The Evolution of Agency and Other Essays*

William S. Cooper   *The Evolution of Reason*

Continued after the index

Evolutionary Ethics and Contemporary Biology

Edited by

GIOVANNI BONIOLO
IFOM Milano

GABRIELE DE ANNA
Università di Udine

CAMBRIDGE UNIVERSITY PRESS
Cambridge, New York, Melbourne, Madrid, Cape Town, Singapore,
São Paulo, Delhi, Dubai, Tokyo

Cambridge University Press
32 Avenue of the Americas, New York, NY 10013-2473, USA

www.cambridge.org
Information on this title: www.cambridge.org/9780521122702

First published 2006
This digitally printed version 2009

A catalog record for this publication is available from the British Library

Library of Congress Cataloging in Publication data

Evolutionary ethics and contemporary biology / edited by Giovanni Boniolo,
Gabriele De Anna.
  p. cm. – (Cambridge studies in philosophy and biology)
Includes bibliographical references and index.
ISBN 0-521-85629-9 (hardback)
1. Ethics, Evolutionary. 2. Bioethics. I. Boniolo, Giovanni.
II. De Anna, Gabriele, 1970– III. Title. IV. Series.
BJ1311.E97 2006
171′.7–dc22   2005028954

ISBN 978-0-521-85629-4 Hardback
ISBN 978-0-521-12270-2 Paperback

Contents

Contents

Contributors

Francisco Ayala, *University of California, Irvine*

Giovanni Boniolo, *Firc Institute for Molecular Oncology (IFOM), Milan*

Stefano Canali, *University of Cassino*

Gabriele De Anna, *University of Udine*

Aldo Fasolo, *University of Torino*

Philip Kitcher, *Columbia University*

Christopher Lang, *University of Wisconsin, Madison*

Danilo Mainardi, *University of Venezia*

Paola Palanza, *University of Parma*

Luca Pani, *Institute of Neuropharmacology and Neurogenetics of the Italian National Research Council, Cagliari*

Stefano Parmigiani, *University of Parma*

Alex Rosenberg, *Duke University*

Michael Ruse, *Florida State University, Tallahassee*

Elliott Sober, *University of Wisconsin, Madison*

Karen Strier, *University of Wisconsin, Madison*

Paolo Vezzoni, *Institute of Biomedical Technologies of the Italian National Research Council, Segrate*

Evolutionary Ethics and Contemporary Biology

Introduction

GIOVANNI BONIOLO AND GABRIELE DE ANNA

A long-standing interest in how biological evolution and ethics relate to each other has focused on the relevance of evolutionism (and subsequent naturalism) to the existence and status of moral values and to the character of moral agency. Discussions regarding the relevance of biology to ethics date back to Aristotle, and when the concept of evolutionism first appeared, it was immediately taken to have important bearings on moral thinking (Maienschein and Ruse 1999). Yet, it was not until the end of the nineteenth century that the interest in these issues really bloomed. More recently, the explanatory and academic success of the "new" biological sciences, such as molecular genetics, ethology, neurobiology, and neuropsychology, opened up promising possibilities for a more profound comprehension of human behavior, including normatively guided agency. Moreover, current debates seem to show that only an integrated contribution of all these sciences can shed light on human agency. Thus, philosophers are now becoming increasingly interested in questions such as whether and how ethics relates to our biological nature, and whether and how aspects of human biology bear upon our social practices.

Within these debates, increasingly pressing questions concern the relevance of recent developments in contemporary biological sciences in furthering our understanding of the relationship between evolutionism and ethics. The problem then arises of understanding how these issues can be correctly framed in philosophical terms. This collection of original essays offers a cutting-edge coverage of the topic and suggests some possible answers.

An attempt to offer a philosophical framework to discuss the idea that evolutionism may be relevant for ethics must face two problems: it should

Gabriele De Anna did part of the work for this book while holding a Visiting Fellowship at the Center for Philosophy of Science, University of Pittsburgh. He is deeply grateful to the Center for hospitality, support, and invaluable philosophical exchange.

1

highlight what sort of contribution evolutionism offers to ethics; and it should characterize precisely what ethics is, that is, what kind of behavior can be called ethical behavior.

Concerning the first issue, there are two main sorts of contribution that evolutionism can offer to ethics. On the one hand, evolutionarily oriented biological sciences can be employed to develop an *explanation* of ethics, in particular, an account of the reasons why ethics exists and has the features it does. On the other hand, biological sciences can be used to offer a *justification* of ethics, namely an account of the reasons why ethical statements are normative and should be followed.

All views affirming that biological sciences have a role in explaining and/or justifying ethics can be broadly called naturalistic. But naturalism comes in different degrees, depending on what the contribution of sciences is claimed to be, and on whether that contribution involves only justification, only explanation, or both. For example, a weak form of naturalism sustains the idea that science is one of the many useful sources to explain and/or justify ethics but accepts that the latter cannot be reduced in any sense to the former. A second, stronger form of naturalism supports the view that science can explain ethics but cannot justify moral discourse. A third, even stronger approach maintains that naturalism concerns the possibility that ethical normativity can be fully explained and justified scientifically. Thus, moral discourse can be reduced to scientific discourse, yet still preserve its colloquial autonomy.

This threefold distinction between different degrees of naturalism is not exhaustive, because different views on the relationship between justification and explanation can complicate the issue. For example, one may contend that, once science has explained all the facts concerning human behavior, including human sentiments and the sense of duty, it is impossible or unnecessary to justify ethics. Ethical discourse will then be viewed as merely fictitious. It seems, in this context, two different lines are possible: an extreme form of naturalism, which sustains that moral discourse has to be eliminated; or a less extreme form, which maintains that moral discourse cannot be renounced, although it is merely illusory.

Let us now turn to the second problem in offering a philosophical framework. What is ethics? What is the object that we hope to explain and/or justify through biological sciences? From a scientific standpoint, ethics can be seen as a particular sort of behavior, typically exhibited by humans, that involves the consideration of norms. Probably the best way to frame this kind of behavior is the multilayered account of human agency recently proposed by Rottschaefer (1998). His model of moral behavior comprises four levels: first, a level of minimally cognitive moral capacities, which are biologically and psychologically founded; second, a behavioral level, involving cognitively acquired moral beliefs and desires, which give rise to moral behavior; third, a reflexive level, concerning moral beliefs and desires about second-level behavioral beliefs and desires, that is, moral norms; and fourth, a self-referential reflexive level, in which a moral agent conceives of himself as a moral agent.

Rottschaefer's analysis suggests that human moral behavior involves two main objects, which biological sciences should help us explain and/or justify: the set of cognitive and emotional traits, which is needed in the four levels and which constitutes what we could call the "human moral capacity"; and the sets of rules or norms (to be empirically identified), which are employed in human agency at levels three and four and which we call "ethical systems."

This distinction further complicates the possible shades of naturalism, because it combines with the justification-explanation distinction and generates a matrix of possibilities. Indeed, one may think that evolutionism may explain and/or justify either the human moral capacity only, ethical systems only, or both.

The chapters in this volume touch upon these two main issues. Chapters in the first part deal with the justificatory and explanatory possibilities (about ethics) of evolutionism. The second part concentrates on some methodological aspects that are central for all attempts to explain and justify ethics. The third part focuses on how recent findings in various biological sciences may help explain the human moral capacity. The fourth part focuses on how recent scientific results may contribute to the explanation of moral systems.

Most of the authors agree with a weak form of naturalism, which claims that evolutionism fails to justify ethics, although it may explain some features of it (which is why the third and the fourth parts are entirely devoted to the problem of explanation). Some suggest an even weaker form of naturalism. According to Boniolo, for example, evolutionism may explain the enabling conditions for the human moral capacity but not ethical systems. Canali, De Anna, and Pani, to offer another example, think that evolutionism can certainly contribute to the explanation of some human ethically relevant cognitive capacity but only in conjunction with other nonbiological considerations.

In the opening chapter, Michael Ruse discusses the main metaethical question raised by evolutionary ethics: can evolutionism justify morality? He distinguishes this metaethical question from the problem of normative morality, that is, what norms we ought to follow. He contends that normative morality can be successfully explained by evolutionism and that this explanation is a matter of empirical results coming from different sciences. When it comes to justify our substantive moral norms (i.e., to point out their foundations, the things that make them compulsory, the reasons why we have to follow them), the concern becomes metaethical and is a matter of philosophical (rather than empirical) investigation. According to Ruse, on the metaethical level evolutionism leads to skepticism, which entails a form of metaethical antirealism. Evolutionism highlights that there are no foundations for our moral norms, because they are the mere result of our evolutionarily originated moral sentiments. Against sociobiologists, Ruse contends that evolutionism cannot offer a justification for normative morality, because evolution itself is not a value, nor does it follow a direction that might be evaluatively qualified as progressive. Evolution, indeed, could have taken a different direction, and we could have ended up with different moral sentiments and, thus, with a different normative morality. Hence, there is no basis for the claim that our moral norms have to be followed. This is not to deny the grip of normative morality on us. The point is that this grip is the mere result of an illusion. Against traditional twentieth-century forms of emotivism and perspectivism, though, Ruse contends that the illusion concerns also the very objectivity of the contents of moral norms. Furthermore, the illusion cannot vanish, because it depends on moral sentiments that were entrenched in our psychology by evolution. The resulting view is a Humean form of moral sentimentalism combined with Darwinian evolutionism.

Giovanni Boniolo, in the second chapter, seems to take a similar stand on the metaethical level, because he claims that moral behaviors are totally judgment-dependent and that certain kinds of relativism must be accepted, even if "anything goes" forms of relativism should be rejected (e.g., epistemological relativism must be accepted, but existential relativism should be rejected). These views seem to recall Ruse's contention that moral judgments are illusory and that particular moral systems cannot be justified by evolution (and, thus, cannot be justified at all), even if we cannot avoid being in the grip of one of them. However, Boniolo seems to disagree with Ruse on the hopes for an evolutionary explanation of ethics. Whereas Ruse subscribes to his long-standing defense of a naturalistic explanation of normative ethics, Boniolo claims that the evolutionary approach can explain the genesis of the enabling conditions for the human moral capacity but not the diversity of ethical systems. Ruse's version of naturalism, thus, is rather weak. Boniolo introduces the distinction among *behavior, moral judgment on behavior,* and *moral capacity* (i.e., the capacity for both formulating and applying moral judgments on behavior, and for acting accordingly). By starting from this distinction, he shows that Darwin had both a theory of the genesis of the moral capacity and a theory of the genesis of the different moral judgments. Only the moral capacity can be naturalized via evolutionary biology, while moral

4

judgments and moral systems cannot, even if some contemporary authors suggest such a possibility. Boniolo's aim is to offer an explanation of moral capacities that follows directly from Darwin's theory of evolution and is compatible with a form of moral (phylo)genetic relativism. As a conclusion, he states that moral capacities have to be considered as an accidental evolutionary outcome that was made possible by the evolution of suitable cerebral-mental traits. Contrariwise, the formulation and the application of moral judgments are purely matters of human culture, which cannot be explained by biological sciences.

Attempts to explain and/or justify ethics assume that humans (and their behavior) can be the object of scientific considerations from an evolutionary perspective, just as all other living beings. But can humans be considered in this way? And, if so, to what extent? This issue is discussed in the chapters of the second part of this volume.

Whether and to what extent humans are part of nature is discussed by Christopher Lang, Elliott Sober, and Karen Strier. Obviously, in order to discuss the relations between biology (which offers a representation of nature) and human beings, we must be extremely clear whether, and in which sense, humans belong to nature. This is particularly relevant when ethics is concerned: evolutionary explanations of ethics can be accepted only if humans and their activities (including ethically guided behavior) are parts of nature in a relevant sense. In developing this point, the authors distinguish between unified and disunified explanations of human features. Unified explanations seek to situate the traits of human beings in a causal framework that can also explain the trait values found in nonhuman species. Disunified explanations claim that the traits of human beings are due to causal processes that are not at work in the rest of nature. The chapter outlines a methodology for testing hypotheses of these two types and draws implications concerning evolutionary psychology, adaptationism, and antiadaptationism. The suggested methodology does not concern moral behavior exclusively but also has fundamental consequences for evolutionary ethics, as the authors recognize.

Besides establishing in which sense human beings are part of nature, investigators must face another extremely relevant methodological aspect: what is the real value of the comparisons between human and nonhuman animals? Although most discussions on the evolutionary status of moral capacities and moral systems are grounded on comparative analyses of human and nonhuman behaviors, are such analyses really well grounded? This topic, which is fundamental for a better comprehension of the relations between evolutionism and ethics, is discussed in the chapter by the neurobiologist Aldo Fasolo. Explanations of human cognitive capacities (including moral capacities) often rely on analogies with capacities of other animals. Such analogies, however, need to meet precise methodological constraints. To what extent are humans similar to other animals? To what extent can we apply to humans models that we have developed for other animals? In contemporary comparative neurobiology, homology is fundamental for any attempt to offer a neuro-biological explanation of human behavior. Fasolo offers criteria that may be useful in distinguishing genuine biological correspondences from loose metaphorical representations in descriptions of human behavior proposed by evolutionary ethicists.

With the fifth chapter, by Giovanni Boniolo and Paolo Vezzoni, we enter the third part of the volume, which focuses on genetic and evolutionary explanations of the human moral capacity. Boniolo and Vezzoni, starting from the antireductionist claim that not everything is in our genes, argue that we cannot overlook that *something* is in our genes. The problem is to understand what that is and to what extent it can constrain our moral capacity. Therefore they investigate in which sense, in some deviant cases, an agent's moral capacity is genetically influenced. Nevertheless, they do not support the idea that genetics morally assesses these deviant cases: genetics does not at all offer the grounds for any moral judgment. They conclude that even if we know, by studying monogenic and polygenic diseases, that our genes, in particular their deviant forms, influence our moral capacity by acting on its enabling conditions, there is not enough scientific ground yet to state in which degree these influences occur.

The sixth chapter, by Stefano Canali, Gabriele De Anna, and Luca Pani, also deals with abnormalities of human behavior. The authors discuss the relevance of evolutionary considerations for psychiatric diagnosis and treatment: psychiatry cannot renounce the notion of the normal functioning of human beings, because the very requirement of a treatment presupposes that the situation to be treated is abnormal and needs to be normalized. However, evolutionary considerations show that the "normal functioning" of a human being is a notion that needs to be tailored to each individual. Generalizations on what is a normal human being are needed, but they subsequently need to be readjusted in light of the particular (genetic and environmental) situation of each individual. When these genetic and environmental situations are considered, kinds of behavior that would otherwise seem pathological may turn out to have an evolutionary significance and thus to make the individual more apt to its environment. Therefore, evolutionary-based psychiatry can help to determine what kinds of human behavior are normal and what kinds are abnormal. In this way, it can help *explain* what the human moral capacity is: that is, what particular cognitive and emotional trait must characterize a normal human being having a complete mastery of his moral agency. (In passing, the authors also hint at a possible line to be taken if one wants to use the notion of human function also to *justify* ethical behavior.) Although their view is in a sense naturalistic, the form of naturalism suggested by the authors is quite weak: in their view, evolutionarily based neurological and psychiatric considerations can provide only some of the considerations that need to be taken into account when determining what normal human behavior and the moral capacity are. In other words, human behavior and moral capacity can only be partly explained by evolutionary psychiatry.

The essay by Parmigiani, De Anna, Mainardi, and Palanza, the seventh chapter of the volume, considers the contribution of ethology to the explanation of ethics. The authors contend that ethological considerations clearly show that several ethically relevant sorts of behavior do not depend entirely on culture but have strong inherited bases. Indeed, several emotional and cognitive traits, which lead to certain sorts of behavior, clearly represent a universal human heritage and have an evident evolutionary significance, because they follow the "selfish gene" pattern of evolution. The authors consider the cases of infanticide and of male jealousy for females. They are so widespread among human populations that it makes sense to speak about a "universal human nature." The authors, however, consider some examples of ethical systems that do not conform to the selfish-gene pattern of evolution or prescribe the kinds of behavior that should be expected from our emotional and cognitive traits (e.g., most contemporary ethical systems claim that infanticide is wrong). This suggests that considerations based on evolutionary ethology can explain the origin of certain human emotional and cognitive traits but cannot explain the origin of moral systems, which depend on culture. Although there may be selective pressures on cultures, the evolution of moral systems does not seem to follow the same patterns of biological evolution. In this way, the authors confirm a conclusion already supported in other chapters included in this collection (e.g., Boniolo's), even if they suggest that biological evolution and cultural evolution may constitute a continuum.

The eighth chapter, by Francisco Ayala, opens the last part of the volume. Ayala tries to offer an account of ethics that can preserve both the not fully naturalistic outlook of sets of norms that he previously proposed (Ayala 1995) and the need for an evolutionary explanation of the reasons why certain sets of norms developed. Ayala suggests that in humans there are two kinds of heredity: the biological and the cultural. Biological inheritance is based on the transmission of genetic information from parents to offspring, very much the same in humans as in other sexually reproducing organisms. Cultural inheritance, on the other hand, is distinctively human, based on transmission of information through a teaching and learning process, which is, in principle, independent of biological parentage. Cultural inheritance makes possible the cumulative transmission of experience from generation to generation. Ayala claims that cultural heredity is a swifter and more effective mode of adaptation to the environment than the biological mode because it can be designed. The advent of cultural heredity ushered in cultural evolution, which transcends biological evolution. The chapter ascertains the causal connections between human ethics and human biology. Ayala's conclusions are that the proclivity to make ethical judgments, that is, to evaluate actions as either good or evil, is rooted in our biological nature, a necessary outcome of our exalted intelligence. On the other hand, the moral codes that guide our evaluations of actions are products of culture, including social and religious traditions. This second conclusion contradicts evolutionists and sociobiologists who claim that the moral good is simply that which is promoted by the process of biological evolution. Ayala thus rejects strong forms of naturalism about the evolutionary explanation of moral systems. Moral codes and, hence, our self-referential reflexive moral understanding are the result of cultural heredity. In this way, cultural evolutionism can hope to explain ethical systems. A justification of ethical systems, though, is still needed.

Philip Kitcher, in the ninth chapter, offers an explanation of the emergence of ethical systems based on ethological information regarding altruism among primates and on considerations concerning the adaptive advantages of the reinforcement of altruistic behavior. Primitive hominids probably lived in social groups rather like those of contemporary chimpanzees and bonobos, groups in which fragile altruistic dispositions were often overridden and in which peacemaking strategies were constantly needed. Kitcher suggests that we can understand the emergence of morality in terms of an ability to reinforce these altruistic dispositions, and that this made it possible to evolve both larger group sizes and a richer array of cooperative projects. He explores this suggestion in the context of what we know about human evolution and about the moral systems that first appeared in the historical record. Altruism is one of the most intensely discussed topics in the literature on evolutionary ethics, to which Kitcher has already offered fundamental contributions. The reasons for the popularity of altruism certainly lie in the fact that it is the most obvious example of a subject that can be studied both in animal groups and in most human moral systems. However, altruism is a difficult starting point for evolutionary ethics, because it does not normally allow one to see a clear and empirically supported line of development toward a full explanation of human morality. In this respect, Kitcher's chapter offers an original contribution, in that his model of the reinforcement of altruistic dispositions opens up the possibility of finding a full-fledged explanation of ethical behavior.

It is worth noting that Kitcher's conclusions contrast with those reached by Parmigiani et al. We then face the problem whether evolutionism can explain ethical systems as well, or only the human moral capacity. Kitcher's considerations on altruism suggest the former, whereas Parmigiani et al.'s considerations on infanticide the latter. This opens a set of questions that this collection cannot settle, but which ethologists and philosophers should discuss. In order to settle the dispute, we need to understand whether the evolutionary advantage of altruism is greater than the evolutionary damage of renouncing infanticide.

In the tenth and final chapter, Alex Rosenberg discusses some of the empirical evidence already discussed by Boniolo and Vezzoni and offers an alternative and novel way of making a philosophical use of the relevant variation correspondences. He suggests that comparative gene sequencing is the only possible source of evidence that could change the interesting and imaginative "just-so stories" of evolutionary game theory into a scientifically confirmed chronology of how cooperation, altruism, sociality, and moral conduct evolved among humans. He argues that prospects for some illumination from this source are not negligible, given the advances in the sequencing of ancient DNA and comparative genomics with our chimpanzee cousins. Rosenberg's comprehensive chapter is an ideal conclusion for the volume, because it shows how different sciences can be integrated in an account of the origin and nature of most of those human capacities which are involved in moral behavior, at all four levels of moral agency. The question still remains whether that account not only explains but also justifies moral norms.

As a final remark, we hope that this collection of essays offers a fully comprehensive and up-to-date picture of the philosophical problems concerning the relations between ethics, evolutionism, and contemporary biological sciences. The collection suggests that moral discourse cannot be eliminated, and that even a mere reduction of morality to sciences is highly problematic. Nonetheless, it shows that scientific findings are relevant for our understanding of all aspects of morality, both the issues concerning our moral capacities associated with the lower levels of the suggested model of moral agency and the higher levels related to our everyday understanding of moral obligations and our moral self-conception. But there is a further, maybe more important reason for which this collection may be useful. The present essays offer methodological reflections on, and actual examples of, the ways in which scientific findings can be used as evidences for a philosophical explanation of human moral behavior. We hope that this will benefit both philosophers and scientists. Philosophers of all persuasions, not only naturalistically minded philosophers, might well see that scientific findings can be usefully adopted in their work on ethics, without the risk of introducing (potentially question-begging) heavy naturalistic assumptions that might lead to a deflationary conception of ethics. Scientists might notice how problematic and moot are the philosophical bearings of their results and may appreciate what sorts of empirical evidence is expected from their work for philosophical purposes.

Some may question the scope of the volume because most of the chapters focus on the evolutionary *explanation* of ethics. The problem of justification, indeed, seems to have been dismissed in the first two chapters on purely skeptical grounds. In truth, the issue of justification remains an open question, and, although most of the authors seem to agree that evolutionism cannot justify ethics, it cannot be prima facie excluded that nonevolutionary, nonskeptical justifications of ethics may be coherent with the explanations advanced here. Moreover, it may even be the case that these explanations can be employed in some attempts of justification that do not rely on evolutionism but are naturalistic nonetheless. For example, recent natural-law attempts to justify ethics – such as those by Philippa Foot (2001) and Mark Murphy (2001), which rely on facts concerning human nature as reasons for actions that may justify ethical systems – may find in the explanations of human moral capacities and ethical systems here presented important insights for the understanding of human nature. It is our hope that these essays may also be of some interest to those working on justificatory projects of this kind.

REFERENCES

Ayala, F. 1995. The Difference of Being Human: Ethical Behavior as an Evolutionary Byproduct. In *Biology, Ethics, and the Origin of Life*, ed. H. Rolston III, 113–135. Boston and London: Jones and Bartlett.

Foot, P. 2001. *Natural Goodness*. Oxford: Oxford University Press.

Maienschein, J., and Ruse, M. 1999. *Biology and the Foundations of Ethics*. Cambridge: Cambridge University Press.

Murphy, M. C. 2001. *Natural Law and Practical Rationality*. Cambridge: Cambridge University Press.

Rottschaefer, W. A. 1998. *The Biology and Psychology of Moral Agency*. Cambridge: Cambridge University Press.

I

The Limits of Evolutionary Explanations and Justifications of Ethics

1

Is Darwinian Metaethics Possible (And If It Is, Is It Well Taken)?

MICHAEL RUSE

> Ethics is an illusion put in place by natural selection to make us good
> cooperators.
>
> Ruse and Wilson 1985

When I first started doing philosophy some forty years ago, evolutionary
ethics was the philosophical equivalent of a bad smell. One knew that not
only was it false, but somehow it was unclean – it was the sign that one had
a tin ear for philosophy. Had not G. E. Moore in *Principia Ethica* shown that
evolutionary ethics commits the greatest of all sins, that it ignores or plows
through the "naturalistic fallacy"? Or, to put matters in a more historical
context, did not evolutionary ethics violate the distinction drawn by David
Hume between "is" and "ought"? Now, however, we have had something of a
sea change, and it is almost the norm for philosophers interested in morality
to admit, with greater or lesser enthusiasm, that evolution surely counts for
something. But how much is that "something"? That is still the matter of
debate.

NORMATIVE ETHICS

In dealing philosophically with morality, there are always two levels to be
discussed: *normative* or *substantive ethics*, which deals with what one ought to
do ("love your neighbor as yourself"), and *metaethics*, which deals with why
one ought to do what one ought to do ("God wants you to love your neighbor
as yourself"). If one is trying to link evolution and normative ethics, then

In this chapter, I have drawn on some of my thinking from my Gifford Lectures given in Glasgow
in the fall of 2001.

13

most obviously one will be trying to show that human ethical relationships are produced by evolution. Clearly, by its very nature, this is a naturalistic process – one is trying to show how people feel about moral statements. One is not judging the moral statements as such, although such an approach does not preclude any argumentation about content. One could get into discussion about such issues as consistency, as well as the relevance of factual claims to moral issues. For instance, one might ask whether one is consistent in opposing capital punishment yet at the same time allowing abortion on request. One might ask whether peace is more likely if one goes to war with Iraq or if one tries other methods of containment. But, ultimately, I take it that one is in the business of description and scientific explanation.

There has been much work done in the past twenty years trying to show how Darwinism does explain (in the sense of showing the origins of) normative ethics (e.g., Sober and Wilson 1997; Wright 1994; Gibbard 1990; Skyrms 1998). Although it is a dirty word in philosophical circles, the key breakthrough was the rise of *sociobiology* in the 1970s, with the various models of kin selection and reciprocal altruism and the like, showing how Darwinian advantage could be gained by helping others (Ruse 1985) – all a kind of enlightened self-interest on the part of the genes. "You scratch my back and I'll scratch yours." Uncomfortable with the "selfish-gene" approach, in recent years a number of holistic-type thinkers have been trying to promote an understanding of selection that emphasizes adaptations for the group (as against adaptations for the individual). I myself am not very keen on this way of seeing things, but here I will not dispute it. The main point is one of overlap. All are attempting to explain normative ethics as the result of evolutionary processes, and by this is meant that natural selection of some kind is the chief causal force. The late Stephen Jay Gould (2002) argued that perhaps mental attributes, and these would presumably include mental moral attributes, are simply what he called "spandrels," that is, by-products of the evolutionary process without any adaptive value. Although there are certainly philosophers who would be sympathetic to Gould's approach, the people who have tried to understand ethics in terms of evolution would dispute this.

I have myself for at least two decades been arguing for such a naturalistic, evolution-based approach to normative ethics (Ruse 1986b, 1995, 2001b). Here I do not intend to retread that material. Frankly, I think there is only so far that a philosopher like myself can take the discussion. A naturalistic approach means just that – one puts oneself in the hands of the scientists. These would include primatologists, students of comparative cultures, game theorists, evolutionary psychologists, economists perhaps, and others. All I will say here is that I find the results thus far very encouraging, although I am sure my critics would say that they would hardly expect me to find otherwise. What I do want to do now rather is to turn to the purely philosophical part of the equation, namely that of justification. What of Darwinian metaethics?

METAETHICS

There is still some hesitation by philosophers on this one. It is one thing to turn normative ethics over to the empiricists. It is quite another to think that the results of empirical science can truly answer questions that are so fundamentally philosophical – so dear to the hearts of those of us who stand in the tradition of Plato and Aquinas and Kant. This ambivalence is shown in a recent piece by the well-known philosopher Philip Kitcher. He asks the question: "So what exactly is the relationship between evolutionary theory and ethics?" Then he gives a preliminary answer.

> Let's start with a simple answer. There are many different projects relating evolutionary biology to ethics, some of which are perfectly sensible, others flawed. The hyper-Darwinian ambition is to show how our understanding of the history yields new basic moral principles. Somewhat less ambitiously, one might contend that Darwinism supports some distinctive metaethical view, that it shows, for example, that moral judgements cannot have truth-values or that moral knowledge is impossible. Much more modestly, we can see the evolutionary understanding of our species as relevant to the tracing of all aspects of human history, including the history of our morality and social systems. Finally, one might suppose that recognition of the kinship of life, coupled with moral principles we already hold, enables us to arrive at new derivative moral judgements – perhaps we come to understand ourselves as having obligations not to treat other animals in particular ways. The simple answer proposes that the first two of these ventures are illegitimate, while the latter two are well grounded. (Kitcher 2003, pp. 411–412)

Kitcher argues that this simple answer is only three-quarters right. The second part of the answer may well be false. "What is more problematic – and more interesting – is the claim about the irrelevance of Darwin for metaethics." All well and good. But do not get too excited. Before he is finished, Kitcher escapes making any definite decisions, concealed as he is in a cloud of apparently judicious hesitation about making any final judgments before all the facts are in.

> In outline, we can view morality as a human phenomenon that enters our history as a device for regulating the conflict between our sympathetic and selfish

dispositions (where regulation plays a key role in the maintenance of our societies) and is further articulated through interactions among different social groups and members' reflections on those interactions. What status this assigns to our moral claims depends, I suggest, on the details of the story, and the details require much more research in evolutionary biology, anthropology, psychology and history than anyone has yet attempted. (p. 415)

Positions of this kind are not unknown in the philosophical community. If the science turns up trumps, I was there before you. And if not, then don't blame me. Run with the hare of naturalism, and hunt with the hounds of antinaturalism – and blame science for your ambivalence.

Let me rush in where angels fear to tread. There is another philosophical tradition to ethics – that of Aristotle, Hobbes, and Hume, where the natural world is considered relevant, all the way down (or up). I believe we do now have enough material to make some judgments and decisions at the metaethical level, and in this discussion I am going to show you why I believe this. I agree that we do not have everything that we would like at the normative level. All of the details – perhaps even the broad strokes – of the natural development of morality have not been explicated and explained. But, as Kitcher himself agrees in the quotation just given, we do have something. Biology – let us now agree for the sake of argument, natural selection – has played some significant role in making us moral beings. Morality is an adaptation like hands and teeth and penises and vaginas. Obviously biology does not play the only role, and we must certainly allow culture some significant part also. How significant we can leave more or less open, between two false extremes – that everything is basically cultural (the blank slate hypothesis) and that everything is basically biological (the genetic determinism hypothesis). The point is that morality has come through human evolution and it is adaptive.

SOCIAL DARWINISM

There is a traditional way of relating evolution and morality, thinking now about metaethical issues – that is, about issues cantering on the justification or foundation of morality. (Why should I do that which I should do?) This is the way of the social Darwinian (Ruse 1996; Richards 1987). Take as a paradigm the nineteenth-century philosopher Herbert Spencer. He argued from the way that things have been to the way that things ought to be. One ferrets out the nature of the evolutionary process – the mechanism or cause of

16

evolution – and then one transfers it to the human realm (if this has not already been done), arguing that which holds as a matter of fact among organisms holds as a matter of obligation among humans (Ruse 1986a, b). Spencer himself started with the struggle for existence and the consequent selective effects: a connection that he made in print in 1852, years after Darwin made the connection but years before Darwin published. He then transferred to the human realm: not much to do here, actually, since Spencer speculated on selective effects showing themselves in the different natures and behaviors of the Irish and the Scots. He concluded that struggle and selection in society translates into extreme laissez-faire socioeconomics: the state should stay out of the way of people pursuing their own self-interests and should not at all attempt to regulate practices or redress imbalances or unfairnesses. Libertarian license therefore is not only the way that things are but the way that they should be.

In fact, Spencer was far from convinced that mid-Victorian Britain was a laissez-faire society, but this is what he hoped fervently it would become.

> We must call those spurious philanthropists, who, to prevent present misery, would entail greater misery upon future generations. All defenders of a Poor Law must, however, be classed among such. That rigorous necessity which, when allowed to act on them, becomes so sharp a spur to the lazy and so strong a bridle to the random, these pauper's friends would repeal, because of the wailing it here and there produces. Blind to the fact that under the natural order of things, society is constantly excreting its unhealthy, imbecile, slow, vacillating, faithless members, these unthinking, though well-meaning, men advocate an interference which not only stops the purifying process but even increases the vitiation – absolutely encourages the multiplication of the reckless and incompetent by offering them an unfailing provision, and *discourages* the multiplication of the competent and provident by heightening the prospective difficulty of maintaining a family. (Spencer 1851, pp. 323–324)

Spencer could sound positively brutal about those who would help the unfortunate within society: "Besides an habitual neglect of the fact that the quality of a society is physically lowered by the artificial preservation of its feeblest members, there is an habitual neglect of the fact that the quality of a society is lowered morally and intellectually, by the artificial preservation of those who are least able to take care of themselves. . . . For if the unworthy are helped to increase, by shielding them from that mortality which their unworthiness would naturally entail, the effect is to produce, generation after generation, a greater unworthiness" (Richards 1987, p. 303).

Michael Ruse

PROGRESS

But how does one justify this move? It is here that Moore and others found the fallacy. Because things are this way, it does not follow that things should be this way. In fact, I myself agree with this criticism, but my experience is that social Darwinians (these days they tend not to be called by this name) find this criticism supremely unimpressive. My sometime coauthor Edward O. Wilson points out that while it is indeed true that one is going from "is" to "ought" – in his own case he is concerned to promote biodiversity, as one does when saving the Brazilian rain forests, and hence goes from the premise that humans need biodiversity to the conclusion that we should promote biodiversity – this in itself hardly makes the inference fallacious (Wilson 1984, 1992, 1994). In science, one is always going from talk of one kind to talk of another kind, and no one thinks this fallacious in itself. In gas theory one goes from talk of molecules bouncing around a chamber at different speeds, to talk of increases in pressure and temperature. Is this any more odd than going from "humans need the forests" to "we ought to preserve the forests"?

We need to dig further into the metaethics of the social Darwinian, and soon the real reason for the confidence becomes apparent. To a person, social Darwinians – call them traditional evolutionary ethicists if you prefer – are progressionists. They think that the course of evolution is upward, from the bad or the nonmoral to the good and the moral and the worthy of value. Hence, to keep this progress going is in itself a good thing. Listen, for instance, to Herbert Spencer (1857). For him, evolution was a transition from the undifferentiated, or what he called the "homogeneous," to the thoroughly mixed up, or what he called the "heterogeneous." Progress was not just a biological or a social phenomenon: it was an all-encompassing world philosophy.

> Now we propose in the first place to show, that this law of organic progress is the law of all progress. Whether it be in the development of the Earth, in the development of Life upon its surface, in the development of Society, of Government, of Manufactures, of Commerce, of Language, Literature, Science, Art, this same evolution of the simple into the complex, through successive differentiations, hold throughout. From the earliest traceable cosmical changes down to the latest results of civilization, we shall find that the transformation of the homogeneous into the heterogeneous, is that in which Progress essentially consists. (p. 3)

Likewise later thinkers of this ilk. Take Edward O. Wilson: "the overall average across the history of life has moved from the simple and few to the more complex and numerous. During the past billion years, animals as a whole

18

evolved upward in body size, feeding and defensive techniques, brain and behavioral complexity, social organization, and precision of environmental control – in each case farther from the nonliving state than their simpler antecedents did" (Wilson 1992, p. 187). He concludes: "Progress, then, is a property of the evolution of life as a whole by almost any conceivable intuitive standard, including the acquisition of goals and intentions in the behavior of animals." The point is made.

And here I think is the reason to be dubious about the metaethics of social Darwinism. Popular though it may be, the very idea of progress in evolution is clouded in problems. It is far from obvious either that natural selection promotes progress or that progress actually occurs, at least in any clear definable and quantifiable way. One can of course label humans as the pinnacle of being – I myself am inclined to do just this – but such an act is arbitrary, at least as applied to evolution. Why not label a dog the pinnacle of being or a buttercup? From a biological point of view, the AIDS virus is far more successful than the gorilla, but does anyone truly want to say that the former is superior in a moral or other value sense than the latter?

In a typically hyperbolic fashion, Stephen Jay Gould (1988) writes: "Progress is a noxious, culturally embedded, untestable, nonoperational, intractable idea that must be replaced if we wish to understand the patterns of history." With respect to human evolution, he writes: "Since dinosaurs were not moving toward markedly larger brains, and since such a prospect may lie outside the capabilities of reptilian design . . . , we must assume that consciousness would not have evolved on our planet if a cosmic catastrophe had not claimed the dinosaurs as victims. In an entirely literal sense, we owe our existence, as large and reasoning mammals, to our lucky stars" (Gould 1989, p. 318). Even if one thinks that this is perhaps a little extreme, there is surely enough truth to make one very wary about biological progress as a basis for one's moral code. Whatever one might say about the normative ethics of social Darwinism – and although I am not very keen on laissez-faire, I am very keen on the rain forests and their preservation – metaethically, the justification seems shaky.

ETHICAL SKEPTICISM

But can one do better? Can one overcome Kitcher's hesitation? I think one can. Remember that, for the sake of argument, we are agreeing – and I think Kitcher gives us this much – that we humans have built in innately, or instinctively if you like, a capacity for working together socially. And this capacity manifests

19

itself at the physical level as a moral sense. Hence, morality or rather a moral sense is something that is hard-wired into humans – mediated and fashioned by culture. Morality has been put there by natural selection in order to get us to work together socially or to cooperate. This is not to say that we do not have freedom in any sense. It is not to say that we never disregard our moral sense, but rather that we do have the moral sense and we have the moral sense not by choice or decision, but because we are human. (Of course, there are going to be psychopaths without a moral sense, but in biology you know that there are going to be exceptions for every rule.) The claim therefore is that when humans find themselves in a position where cooperation might pay, morality kicks into place.

This is not to say that we always will cooperate or be moral. We are influenced by many factors, including selfish and other sorts of desires. But morality is one of these factors, and overall we humans do generally work together. Sometimes the morality backfires. I might go to the aid of a drowning child, and drown myself. This is hardly in my self-interest. But on balance it is in my interests to have the feeling that I ought to help people in distress, particularly children in distress. This is both because I myself was at some stage of my life a child, and also because I myself will probably have or be having children. I want others to be prepared to make a risk on my behalf or on the behalf of my children.

Let it also be stressed that humans have a genuine sense of morality. It is the kind of morality that someone like Immanuel Kant (1949) talks about. This is not a scientific position of pure ethical egoism in the sense that we are all selfish people just simply calculating for our own ends. We are rather people with a real moral sense, a feeling of right and wrong and obligation. Admittedly, at the causal level, this may well be brought about by individual selection maximizing our own reproductive ends. But the point is that, although humans are produced by selfish genes, selfish genes do not necessarily produce selfish people. In fact, selfish people in the literal sense tend to get pushed out of the group or ostracized pretty quickly. They are simply not playing the game. In a way, therefore, we have a kind of social contract. But note that it is not a social contract brought about, in the long-distant past, by a group of gray-bearded, old men sitting around a campfire. It is rather a social contract brought on by our biology, that is to say, by our genes as fashioned and selected by natural selection.

This then is the Darwinian perspective on the evolution and current nature of morality. Let us now see how this plays out when we try to put things into a philosophical context. What kind of metaethical justification can one give for such claims as that one ought to be kind to children and that one ought to favor one's own family over those of others? I would argue, paradoxically but truthfully, that ultimately no justification can be given! That is to say, I argue that at some level one is driven to a kind of moral skepticism: a skepticism, please note, about foundations rather than about substantive dictates. What I am saying therefore is that, properly understood, the Darwinian approach to ethics leads one to a kind of moral nonrealism (Ruse 1986b).

In this respect, the Darwinian metaethics I am putting forward in this chapter differs very dramatically from traditional Darwinian metaethics, that of social Darwinism. There, the foundational appeal is to the very fact of evolution. People like Herbert Spencer and Edward O. Wilson argue that one ought to do certain things because, by so doing, one is promoting the welfare of evolution itself. Specifically, one is promoting human beings as the apotheosis of the evolutionary process – a move condemned by philosophers as a gross instance of the naturalistic fallacy, or as a flagrant violation of Hume's Law (that which denies that one can move legitimately from the way that things are to the way that things ought to be). My kind of evolutionary metaethics agrees with the philosopher that the naturalistic fallacy is a fallacy and so also is the violation of Hume's Law. My kind of evolutionary metaethics also agrees that social Darwinism is guilty as charged. But my kind of evolutionary metaethics takes this failure as a springboard of strength to its own position. The Darwinian metaethics of this chapter avoids fallacy, not so much by denying that fallacy is a fallacy, but by doing an end run around it, as it were. There is no fallacious appeal to evolution as foundations because there are no foundations to appeal to!

OBJECTIFICATION

To be blunt, my Darwinian says that substantive morality is a kind of illusion, put in place by our genes, in order to make us good social cooperators (Ruse and Wilson 1985). I would add that the reason why the illusion is such a successful adaptation is that not only do we believe in substantive morality, but we also believe that substantive morality does have an objective foundation. An important part of the phenomenological experience of substantive ethics is not just that we feel that we ought to do the right and proper thing, but that we feel that we ought to do the right and proper thing because it truly is the right and proper thing. As John Mackie (1979) argued before me, an important part of the moral experience is that we objectify our substantive ethics. There are in fact no foundations, but we believe that there are in some sense foundations.

21

There is a good biological reason why we do this. If, with the emotivists, we thought that morality was just simply a question of emotions without any sanction or justification behind them, then pretty quickly morality would collapse into futility. I might dislike you stealing my money, but ultimately why should you not do so? It is just a question of feelings. But in actual fact, the reason why I dislike you stealing my money is not simply because I do not like to see my money go, but because I think that you have done wrong. You really and truly have done wrong in some objective sense. This gives me and others the authority to criticize you. Substantive morality stays in place as an effective illusion because we think that it is no illusion but the real thing. Thus, I am arguing that the epistemological foundation of evolutionary ethics is a kind of moral nonrealism, but that it is an important part of evolutionary ethics that we think it is a kind of moral realism.

(This is my counter to the worries expressed by people like Alex Rosenberg [2003], who point out that the kind of position that I endorse is close to the twentieth-century moral philosophy of emotivism – where ethical claims are simply emotive utterances – and who point out also that emotivism is clearly false. "Killing babies is wrong" is not just an emotive cry, but a claim about something's being truly really wrong. For me, substantive ethics is only emotion, but it means more than that. Ethics is subjective, but its meaning is objective.)

SPIRITUALISM

In a way, what has been given thus far is just a statement rather than a proof. What justification can I offer for my claim that evolution points toward ethical skepticism (about foundations)? Why should one not say that there truly is a moral reality underlying morality at the substantive level, and that our biology has led us to it. After all we would surely want to say that we are aware of the speeding train bearing down on us because of our biology, but this in no sense denies the reality of the speeding train (Nozick 1981). Why should we not say, in a like fashion, that we are aware of right and wrong because ultimately there is an objective right and wrong lying behind moral intuitions?

However, things are rather different in the moral case from the speeding-train case. A more insightful analogy can be drawn from spiritualism. In the First World War, when so many young men were killed, the bereaved – the parents, the wives, the sweethearts, on both sides of the trenches – often went to spiritualists, hoping to get back in touch with the departed dead. And indeed they would get back in touch. They would hear the messages come through the Ouija boards or whatever assuring them of the happiness of the now deceased. Hence, the people who went to spiritualists would go away comforted. Now, how do we explain this sort of thing? Cases of fraud aside, we would say that people were not listening to the late departed, but rather were hearing voices created by their own imaginations, which were in some sense helping them to compensate for their loss. What we have here is some kind of individual illusion brought about by powerful social circumstances. No one would think that the late Private Higgins was really speaking to his mum and dad. Indeed, there are notorious cases where people were reported killed and then found not to be dead. How embarrassing it would be to have heard the late departed assure you of his well-being and then to find out that the late departed was in fact lying injured in a military field hospital.

In the spiritualism case, once we have got the causal explanation as to why people hear as they do, we recognize that there is no further call for ultimate foundations. I would argue that the biological case is very similar. That there are strong biological reasons for cooperation; naturally, we are going to be selfish people, but as cooperators we need some way to break through this selfishness; and so our biology has given us morality in order to help us do it. Once again I stress that this is not to say that we are always going to be moral people: in fact, we are an ambivalent mixture of good and bad, as the Christian well knows (Ruse 2001a). It is to say that we do have genuine moral sentiments that we think are objective, and that these were put in place by biology. Once we recognize this, we see the sentiments as illusory – although, because we objectify, it is very difficult to recognize this fact. That is why I am fairly confident that my having told you of this fact will not now mean that you will go off and rape and pillage, because you now know that there is no objective morality. The truth does not always set you free.

PROGRESS AGAIN

But still you might protest that does not mean that there is no objective morality behind all of this: either an objective morality of a Platonic ilk that actually exists out there, or an objective morality of the Kantian form that is a kind of necessary condition for rational beings' getting along. Here, however, the Darwinian can come back with a further argument, namely one based on the doubts expressed earlier about biological progress. There is no natural climb upward from the blob to the human, from the monad to the man, as people used to say in the nineteenth century. Rather, evolution is a directionless process, going nowhere rather slowly (Ruse 1993; McShea 1991). What this means in this particular context is that there is really no reason why humans might not have evolved in a very different sort of way, without the kind of moral sentiments that we have. From the Darwinian perspective, there is no ontological compulsion about moral thinking.

It is true that, as Kant stressed, it may possibly be that social animals may necessarily have to have certain formal rules of behavior. But it is not necessarily the case that these formal rules of behavior have to incorporate what we would understand as commonsense (substantive) morality. In particular, we might well have evolved as beings with what I like to call the "John Foster Dulles system of morality," so named after Eisenhower's secretary of state during the Cold War in the 1950s. Dulles hated the Russians, and he knew that the Russians hated him. He felt he had a moral obligation to hate the Russians because, if he did not, everything would come tumbling down. But because there was this mutual dislike, of a real obligation-based kind, there was in fact a level of cooperation and harmony. The world did not break down into war and destruction. As a Darwinian, it is plausible to suggest that humans might have evolved with the John Foster Dulles kind of morality, where the highest ethical calling would not be love your neighbor but hate your neighbor. But remember that your neighbors hate you, and so you had better not harm them because they are going to come straight back at you and do the same.

Now, at the very least, this means that we have the possibility not only of our own (substantive) morality but of an alternative, very different kind of morality: a morality that may have the same formal structure but which certainly has a different content. The question now is, if there is an objective foundation to substantive morality, which of the two is right? At a minimum, we are left with the possibility that we humans now might be behaving in the way that we do but that in fact what is objective morality is something quite else from what we believe. We believe what we do because of our biology, and we believe that because of our biology that our substantive morality is objectively justified. But the true objective morality is something other from what we have.

Obviously, this is a sheer contradiction to what most people mean by objective morality. What most people mean by objective morality incorporates the fact that it is going to be self-revealing to human beings. Not necessarily to all human beings but – like Descartes' clear and distinct ideas – certainly self-revealing to all decent human beings who work hard at it. So, given Darwinism, we have a refutation of the existence of such a morality. Darwinian evolutionary biology is nonprogressive, pointing away from the possibility of our knowing objective morality. We might be completely deceived, and because objective morality could never allow this, it cannot exist. For this reason, I argue strongly that Darwinian evolutionary theory leads one to a moral skepticism, a kind of moral nonrealism.

CONCLUSION

This then is my counter to folk like Philip Kitcher. And if you point out that, far from being very original, my whole position starts to sound very much like that of David Hume, who likewise thought that morality was a matter of psychology rather than reflection of nonnatural objective properties, I shall take this as a compliment, not a criticism. It is indeed true that I regard my position as that of David Hume – brought up to date via the science of Charles Darwin. What better mentors could one have than that?!

REFERENCES

Gibbard, A. 1990. *Wise Choices, Apt Feelings: A Theory of Normative Judgment.* Cambridge, Mass.: Harvard University Press.

Gould, S. J. 1988. On Replacing the Idea of Progress with an Operational Notion of Directionality. In *Evolutionary Progress*, ed. M. H. Nitecki, 319–338. Chicago: University of Chicago Press.

Gould, S. J. 1989. *Wonderful Life: The Burgess Shale and the Nature of History.* New York: W. W. Norton.

Gould, S. J. 2002. *The Structure of Evolutionary Theory.* Cambridge, Mass.: Harvard University Press.

Hume, D. 1978. *A Treatise of Human Nature.* Oxford: Oxford University Press.

Kant, I. 1949. *Critique of Practical Reason.* Chicago: University of Chicago Press.

Kitcher, P. 2003. Giving Darwin His Due. In *Cambridge Companion to Darwin*, ed. J. Hodge and G. Raddick, 399–420. Cambridge: Cambridge University Press.

Mackie, J. 1979. *Hume's Moral Theory.* London: Routledge and Kegan Paul.

McShea, D. W. 1991. Complexity and Evolution: What Everybody Knows. *Biology and Philosophy* 6: 303–325.

Moore, G. E. 1903. *Principia Ethica.* Cambridge: Cambridge University Press.

Nozick, R. 1981. *Philosophical Explanations.* Cambridge, Mass.: Harvard University Press.

Richards, R. J. 1987. *Darwin and the Emergence of Evolutionary Theories of Mind and Behavior.* Chicago: University of Chicago Press.

Rosenberg, A. 2003. Darwinism in Moral Philosophy and Social Theory. *Cambridge Companion to Darwin*, ed. J. Hodge and G. Raddick, 310–332. Cambridge: Cambridge University Press.

Ruse, M. 1985. *Sociobiology: Sense or Nonsense?* Dordrecht: Reidel.

Ruse, M. 1986a. *Taking Darwin Seriously: A Naturalistic Approach to Philosophy.* Oxford: Blackwell.

Ruse, M. 1986b. Moral Philosophy as Applied Science. *Philosophy* 61: 173–192.

Ruse, M. 1993. Evolution and Progress. *Trends in Ecology and Evolution* 8: 55–59.

Ruse, M. 1995. *Evolutionary Naturalism: Selected Essays.* London: Routledge.

Ruse, M. 1996. *Monad to Man: The Concept of Progress in Evolutionary Biology.* Cambridge, Mass.: Harvard University Press.

Ruse, M. 2001a. *Can a Darwinian Be a Christian? The Relationship between Science and Religion.* Cambridge: Cambridge University Press.

Ruse, M. 2001b. Altruism: A Darwinian Naturalist's Perspective. In *Altruism*, ed. J. Schloss. New York: Oxford University Press.

Ruse, M., and Wilson, E. O. 1985. The Evolution of Morality. *New Scientist* 1478: 108–128.

Skyrms, B. 1998. *Evolution of the Social Contract.* Cambridge: Cambridge University Press.

Sober, E., and Wilson, D. S. 1997. *Unto Others: The Evolution of Altruism.* Cambridge, Mass.: Harvard University Press.

Spencer, H. 1851. *Social Statics; or, The Conditions Essential to Human Happiness Specified and the First of Them Developed.* London: J. Chapman.

Spencer, H. 1857. Progress: Its Law and Cause. *Westminster Review* 67: 244–267. Reprinted in *Essays Scientific, Political and Speculative*, 3 vols. (London: Williams and Norgate, 1901).

Wilson, E. O. 1984. *Biophilia.* Cambridge, Mass.: Harvard University Press.

Wilson, E. O. 1992. *The Diversity of Life.* Cambridge, Mass.: Harvard University Press.

Wilson, E. O. 1994. *Naturalist.* Washington, D.C.: Island Books/Shearwater Books.

Wright, R. 1994. *The Moral Animal: Evolutionary Psychology and Everyday Life.* New York: Pantheon.

2

The Descent of Instinct and the Ascent of Ethics

GIOVANNI BONIOLO

> Let us then liken the soul to the natural union of a team of winged
> horses and their charioteer. The gods have horses and charioteers that are
> themselves all good and come from good stock besides, while everyone
> else has a mixture. To begin with, our driver is in charge of a pair of
> horses; second, one of his horses is beautiful and good and from stock
> of the same sort, while the other is the opposite and has the opposite sort
> of bloodline. This means that chariot-driving in our case is inevitably a
> painfully difficult business.
>
> Plato, *Phaedrus*, 246 a–d

MORAL CAPACITY

Reflecting on the biological foundations of human moral behavior within a
Darwinian framework should involve something too often forgotten: finding
actual links with what Darwin proposed in his writings. This task should not
be intended as a philosophically idle form of deference to an author of the
past, but as a necessary step to recover the correct historical and theoretical
bases of the problems we are facing. Indeed, an interest in the history does not
necessarily mean to commit fallacies akin to *argumentum ad verecundiam*, but
it can take the form of a correct *argumentum ad auctoritatem*. In considering
what Darwin wrote about the biological bases of morality, I will recall some
of the theses of *The Origin of Species* and *The Descent of Man* that, *mutatis
mutandis*, could be accepted without much difficulty by those who believe in
the correctness and validity of (neo-)Darwinian evolutionism.

What may be found astonishing by some – in particular, by those who
have never had (or never wanted) the opportunity to venture into Darwin's
fascinating works – is that he not only raised the problem of the biological
roots of ethics but offered a good solution, too. Indeed, his solution is so

27

convincing (even if, of course, it is worked out at the phenotypic level and without contemporary philosophical technicalities) that even in our day it can hardly be opposed.

While expounding Darwin's theses, I have the chance to retrieve a distinction that, though fundamental for the present issue, is almost always neglected – that is, the distinction among *behavior, moral judgment on behavior*, and *moral capacity*, to be intended as the capacity for both formulating and applying moral judgments on behavior, and for behaving accordingly.[1] With this distinction in place, I first contend that Darwin has both a theory of the genesis of human moral capacity (i.e., a theory of what the enabling conditions for the moral capacity are, and of how they have evolved) and a theory, to be kept well distinct from the former, of the genesis of different moral judgments. Then, I suggest that more than one theory of the genesis of the different moral judgments can be coherent with the same theory of the genesis of moral capacity. Finally, I argue that only the latter can involve a naturalization via evolutionary biology, whereas the former cannot, even if some contemporary authors support this possibility.

What I propose might be superficially considered as a form of nihilism, since it will be argued that neither the so-called moral behavior nor moral judgments on behavior have biological *fundamenta inconcussa*. Actually it is not so. What I am suggesting is, on the one hand, an unavoidable consequence of (neo-)Darwinian evolutionism and, on the other hand, an attempt to open up a realm of true morality. For I am claiming that moral responsibility cannot be remitted to a religion, an ideology, or a mysterious *quid* that we simplistically call "nature." In my view, for a (neo-)Darwinian evolutionist there should not be any "theologization of morality," "ideologization of morality," or "naturalization of morality," but only a naturalization (i.e., a *biologization*) of the enabling conditions for moral capacity. Man should be considered fully responsible for his moral judgments and for his behavior, even if his judgments and his behavior depend on his moral capacity, the enabling conditions of which in turn, as Darwin taught us, are entirely dependent on his biological evolution.

From this point of view, Plato's myth of the two horses and the charioteer can be read as if the enabling conditions for the human moral capacity were one of the horses, and moral judgments were the other. Plato's metaphor, then, turns out to suggest that the enabling conditions, due to biological evolution, and moral judgments, due to human affairs and culture, should not

---

[1] To give a precise and uncontroversial definition of moral capacity is not at all easy; cf. Thomasma and Weisstub 2004.

be mixed up willy-nilly, even if a correct analysis of the latter cannot be possible without a good understanding of the former. This desired result can only be accomplished if the charioteer – as Plato wrote – is able to tackle his "painfully difficult business" with competence and ability, without getting himself bewitched by fundamentalist sirens, of whatever kind they might be: religious, ideological, or naturalistic.

Darwin's Theory of the Genesis of the Enabling Conditions for Moral Capacity

In chapter 7 of *The Origin*,[2] Darwin analyzes the relations among the instinctual behavior of lower animals, that of higher animals, and that of man. It would seem proper to start an inquiry of this kind with a definition of "instinct," but Darwin claims that he does not intend to risk such a definition. Nonetheless, he gives it only a few lines later: "An action, which we ourselves should require experience to enable us to perform, when performed by an animal, more especially by a very young one, without experience, and when performed by many individuals in the same way, without their knowing for what purpose, is usually said to be instinctive" (1859, p. 234).

After this clarification, Darwin distinguishes *instinctual behavior* and *habitual behavior*. In his view, the former is a trait inherited by an individual – although he could not know exactly what this meant. The latter, instead, is an unconscious – and sometimes will-independent – behavior that is not a trait of an entire population but only of some individuals. Habitual behavior is the result of a conscious and continuous repetition of a certain noninstinctual behavior. It is a kind of behavior that one acquires by means of training and, from a certain point on, results so naturally that it seems to have one of the characteristics pointed out by Darwin for instinctual behavior: it is "performed . . . without their knowing for what purpose." As a consequence, instinctual behavior and habitual behavior may be so similar that an observer could fail to distinguish them.[3]

It is worth noting that the discussion of this issue is one of the many times in which Darwin's Lamarckism comes to light. According to Darwin, certain sorts of habitual behavior can become hereditary, and therefore instinctual. Nevertheless, Darwin has a clear perception of three different behavioral levels: instinctual behavior; noninstinctual but habitual behavior; and noninstinctual and nonhabitual behavior, which characterizes our species, *Homo*

---

[2] It is chapter 8 in the 1871 edition.
[3] By the way, this question is still open: what is instinctive and what is not?

sapiens. I accept this threefold distinction, even if I prefer to speak of *instinctual behavioral phenotype*; *noninstinctual habitual behavioral phenotype*; and *noninstinctual nonhabitual behavioral phenotype*.

Darwin himself offers an account of behavior that makes it suitable to be called "phenotype." As we can realize by reading both *The Origin* and *The Descent*, behavioral traits, from the point of view of their evolution, must be considered exactly as the physical traits of living beings. Moreover, to call "noninstinctual" a kind of behavioral phenotype seems also acceptable. In addition, it should be noted that it is well known that physical phenotypes are plastic, that is, that there are *norms of reaction*, in virtue of which a given genotype can express different physical phenotypes in different environmental conditions (cf. Futuyma 1979; Ridley 2004). Once the existence of behavioral phenotypic traits is accepted, it seems plausible to conclude that behavioral phenotypes must also be plastic.

With reference to this very close evolutionary parallelism between physical and behavioral phenotypes, in the last chapter of *The Origin* Darwin writes some very clear lines: "we admit the following propositions, namely, – that gradations in the perfection of any organ or instinct, which we may consider, either do now or could have existed, each good of its kind, – that all organs and instincts are, in ever so slight a degree, variable, and, – lastly, that there is a struggle for existence leading to the preservation of each profitable deviation of structure or instinct. The truth of these propositions cannot, I think, be disputed" (1859, p. 435).

Thus, like all physical phenotypes, instinctual phenotypes evolve by means (also) of the environmental selection of small casual mutations.

If in *The Origin* Darwin discusses the evolution of instincts only in one chapter, in *The Descent* he spends many pages extensively discussing the evolutionary steps from lower animals to higher animals and, lastly, to *Homo sapiens*. In this 1871 work, especially in the first part, Darwin argues in the usual way (i.e., through an incredible amount of cases drawn from all kinds of sources) that the shift from lower animals to *Homo sapiens* occurred by means of a gradual evolutionary process of both cerebral-mental traits and instinctual behavioral traits. Naturally, in this case there is no evolutionary gap; also in this case *natura non facit saltus*.

Concerning emotions and basic mental powers (imagination, curiosity, capacity for imitation, reason, attention, memory), Darwin claims that all animals possess them, even if different species possess them at different degrees, with *Homo sapiens* ranking first (Darwin 1871, pp. 66–68). Regarding higher mental powers (abstractive and conceptual capacities, self-consciousness, and mental individuality), Darwin is sure that they belong to higher animals, in particular to primates, but he hesitates about the possibility of assigning them to lower animals. On these grounds, and considering the basic principle that any biological event, or process, must be observed from an evolutionary point of view, he feels confident in drawing the following conclusion: "If these [basic mental] powers, which differ much in different animals, are capable of improvement, there seems no great improbability in more complex faculties, such as the higher forms of abstraction, and self-consciousness, etc., having been evolved through the development and combination of the simpler ones" (1871, p. 86).

Therefore, *Homo sapiens'* species-specific cerebral-mental traits are nothing but the evolutionary result of selective processes applied to the mutations of its ancestors' cerebral-mental traits.

Such an emphasis on the cerebral-mental traits is not accidental. Rather, it is what we need to understand the core of Darwin's solution to the problem of the roots of moral capacity – that is, to the problem concerning its enabling conditions. The possibility of a gradual evolutionary process leading from a certain subclass of instinctual behavior (i.e., social instinctual behavior) to moral behavior depends on a gradual evolutionary process of cerebral-mental traits.

Darwin begins his argumentation by recalling that an evolutionary process leading to the emergence of ethics could only take place in living beings that already possess a particular kind of instinctual phenotype: social instinctual phenotype.[4] That is, a necessary condition for an evolutionary process leading to an ethically governed way of living is that it occurs in animals endowed with social instinctual behavior. Of course, this kind of behavioral phenotype is still an evolutionary result of positive selective processes, connected to preexisting nonsocial instinctual behavioral phenotypes:

> It has often been assumed that animals were in the first place rendered social, and that they feel as a consequence uncomfortable when separated from each other, and comfortable whilst together; but it is a more probable view that these sensations were first developed, in order that those animals which would profit by living in society, should be induced to live together, in the same manner as the sense of hunger and the pleasure of eating were, no doubt, first acquired in order to induce animals to eat. (1871, p. 108)

Nevertheless, even if a social instinctual behavior is necessary, it is not sufficient. For a necessary biological precondition must be satisfied: a suitable

---

[4] At this point, it should be noted that as Darwin – in *The Origin* – did not linger upon the origin of species but started by assuming that there are species, so – in *The Descent* – he did not linger upon the origin of the instinctual phenotypes but assumed that there are instinctual phenotypes.

cerebral-mental evolution must have occurred. In this way, Darwin has all he needs to formulate his solution to the problem of the biological roots of the moral capacity: "The following proposition seems to me in a high degree probable namely, that any animal whatever, endowed with well-marked social instincts, the parental and filial affections being here included, would inevitably acquire a moral sense or conscience, as soon as its intellectual powers had become as well, or nearly as well developed, as in man" (1871, p. 101).

Therefore, according to Darwin, any species that has evolved suitable cerebral-mental traits can arrive at a stage in which its members can both formulate and apply moral judgment and behave accordingly, that is, at a stage in which they are moral agents.

Certainly, a species may evolve in a way such that its members do not judge behavior in the same way as we, humans, do; nevertheless, their moral judgments might be very similar (1871, p.102). Note that Darwin is not discussing the genesis of moral judgments of a certain kind, but the genesis of the enabling conditions for the capacity both for formulating and applying moral judgments and for behaving accordingly.

It should be still observed that such capacity must not be qualified as "moral" in virtue of the fact that it is intrinsically moral. Actually only *a posteriori* can we know that it gives its owners the power both to formulate and to apply moral judgments and to behave accordingly. Therefore, only *a posteriori* are we allowed to call it "moral." This aspect is extremely important, from the point of view of the foundations of ethics, because it leads to an antifundamentalist and antiessentialist position.

Moreover, it should be underlined that in Darwin's view – and this is really a very strong and impressive claim – the human moral capacity does not belong to man because he is a privileged being, in whatever sense this claim might be understood, but only because of casual and nonteleological events that occurred along his phylogenetic evolution. In other words, it is precisely on the basis of the just highlighted phylogenetic casual and ateleological events that we can support the idea that man is what he is. It was the natural history of our species that made us "privileged,"[5] and not some unknown or unknowable metaphysical property gifted by some God.

In the light of the foregoing considerations, we can claim that, according to Darwin, *Homo sapiens'* enabling conditions for the moral capacity are an evolutionary outcome, which should not be necessarily restricted to that species.

---

[5] Of course, it is a relative privilege. In its niche, any species is privileged by its phylogenetic history.

However, why do these enabling conditions concern the evolution of cerebral-mental traits? Darwin's answer is extremely relevant from a biological point of view, and philosophically quite sophisticated:

> As we cannot distinguish between motives [that induce to act in a certain way], we rank all actions of a certain class as moral, if performed by a moral being. (1871, p. 115)

> A moral being is one who is capable of reflecting on his past actions and their motives of approving of some and disapproving of others; and the fact that man is the one being who certainly deserves this designation, is the greatest of all distinctions between him and the lower animals. (1871, p. 633)

That is, morality is not an intrinsic property of behavior but depends on the fact that a certain act is performed by a moral agent, that is, an agent capable of moral judgments on behavior. First, this means that, on the one hand, there are behaviors and, on the other hand, there are moral judgments on behaviors. Second, it means that, in order to both formulate and apply moral judgments on behaviors and to behave accordingly, one must have a suitable cerebral-mental structure.[6]

This claim about the distinction between behavior and moral judgments on behavior (which can be formulated and applied only by a certain kind of animal, in particular those having reached a certain stage of their cerebral-mental evolution) has two important consequences.

First, claiming that certain behaviors are intrinsically moral or immoral is wrong. They can be called moral only insofar as they are thus valued by a living being endowed with the moral capacity in virtue of its cerebral-mental evolutionary history (it has the enabling conditions). That is, behavior is never intrinsically moral or immoral. Morality and immorality are not intrinsic characteristics of behavior, but judgment-dependent properties of behavior. Naturally, as Darwin also admits, "I am aware that the conclusions arrived at in this work will be denounced by some as highly irreligious" (ibid.), but *c'est la vie*, or, rather, *c'est l'évolution de la vie*.

Second, because *Homo sapiens* seems to be the only species that has reached the suitable cerebral-mental evolutionary stage, that is, the only one possessing the suitable enabling conditions, any attempt to analyze the genesis, or the status, of moral theories by comparing nonhuman behavior with human behavior must be looked at with suspicion (cf. Parmigiani et al., Chapter 7 in this volume; Fasolo, Chapter 4 in this volume). Moral capacity,

---

[6] A deep understanding of this claim would require an analysis of both the phylogenetic evolution of the nervous system and mental capacities.

that is, the capacity both for formulating and applying moral judgments, and for behaving accordingly, enters the scene of phylogenesis only after suitable cerebral-mental structures have appeared through casual ateleological processes.

Before going on, it is worth noting that the distinction between behavior and moral judgments on behavior put forward by Darwin was not a novelty. It had already been suggested by several philosophers throughout the history of Western thought, although Darwin was the first to reach that point of view through a biological path. For example, a couple of centuries before Darwin, while theorizing on the phenomenology of behavior and on moral judgments, M. de Montaigne, in his *Essais* (1572–1592), showed that the same sort of behavior might be subject to many different moral judgments. Moreover, only a few years after *The Descent*, while traversing a philosophical-deconstructive route, F. Nietzsche argued in his *Zur Genealogie der Moral* (1887) that each moral judgment has a long story and that, in order to understand its real status, we should know the human and social circumstances from which it emerged.

Of course, neither Montaigne nor Nietzsche focused his attention on the biological roots of the moral capacity. Instead, they tried to show, respectively, the interplay among moral judgments and their genesis in human affairs. Nevertheless, both attempts were grounded on the implicit assumption that there is a deep difference between behavior and moral judgments on it. This is exactly one of the conclusions supported by Darwin and wonderfully described by Villiers de L'Isle-Adam in the first of his *Contes crues*, "Les demoiselles de Bienfilâtre":

> Pascal nous dit qu'au point de vue des faits, le Bien et le Mal sont une question de latitude. En effet, tel acte humain s'appelle crime, ici, bonne action, là-bas, et réciproquement. . . . Les actes sont donc indifférents en tant que physiques: la conscience de chacun les fait, seule, bons ou mauvais. Le point mystérieux qui gît au fond de cet immense malentendu est cette nécessité native où se trouve l'Homme de se créer des distinctions et des scrupules, de s'interdire telle action plutôt que telle autre, selon que le vent de son pays lui aura soufflé celle-ci ou celle-là: l'on dirait, enfin, que l'Humanité tout entière a oublié et cherche à se rappeler, à tâtons, on ne sait quelle Loi perdue. (1874, p. 1)

Darwin's Theory on the Genesis of the Different Moral Judgments

As said, Darwin did not propose only a theory of the biological roots of the moral capacity but also a theory of the genesis of the differences among moral judgments. Even though the two theories must be kept distinct, it is interesting to dwell upon the latter.

In the section entitled "The More Enduring Social Instincts Conquer the Less Persistent Instincts" and in the "Concluding Remarks" in chapter 4 of *The Descent*, Darwin argues that our moral judgments and moral emotions arise because we are embedded in a social environment and are subjected to the approval and disapproval of our likes. Therefore, "Man prompted by his conscience, will through long habit acquire such perfect self-command, that his desires and passions will at last yield instantly and without a struggle to his social sympathies and instincts, including his feeling for the judgment of his fellows" (1871, p. 119). That is, prompted by the "feeling for the judgment of their fellows," men of a certain population must have gradually come to formulate a set of values and rules that nonintentionally and slowly led to moral judgments. As a consequence, "The imperious word ought seems merely to imply the consciousness of the existence of a rule of conduct, however it may have originated. Formerly it must have been often vehemently urged that an insulted gentleman ought to fight a duel. . . . If they fail to do so, they fail in their duty and act wrongly" (ibid.). Thus, moral judgments are nothing but the nonintentional consequences of certain interests of a given human population *to follow* certain kinds of behavior – which are praised – and *not to follow* other kinds – which are blamed:

> The wishes and opinions of the members of the same community, expressed at first orally, but later by writing also, either form the sole guides of our conduct, or greatly reinforce the social instincts; such opinions, however, have sometimes a tendency directly opposed to these instincts. This latter fact is well exemplified by the Law of Honour, that is, the law of the opinion of our equals, and not of all our countrymen. The breach of this law, even when the breach is known to be strictly accordant with true morality, has caused many a man more agony than a real crime. We recognise the same influence in the burning sense of shame which most of us have felt, even after the interval of years, when calling to mind some accidental breach of a trifling, though fixed, rule of etiquette. The judgment of the community will generally be guided by some rude experience of what is best in the long run for all the members; but this judgment will not rarely err from ignorance and weak powers of reasoning. Hence the strangest customs and superstitions, in complete opposition to the true welfare and happiness of mankind, have become all-powerful throughout the world. We see this in the horror felt by a Hindoo who breaks his caste, and in many other such cases. It would be difficult to distinguish between the remorse felt by a Hindoo who has yielded to the temptation of eating unclean food, from that felt after committing a theft; but the former would probably be the more severe. (1871, pp. 125–126)

Thus, a moral judgment is grounded on an interest of a community and on the feeling for blame or for praise.[7]

An upshot of this position is that, in order to avoid blame, one needs not to break rules and must have been properly trained (i.e., educated and self-educated), so that one possesses a suitable self-control and a sufficient command of one's own behavior. Certainly, the development of these moral skills does not require the explicit knowledge of all steps:

> [H]ow so many absurd rules of conduct, as well as so many absurd religious beliefs, have originated, we do not know; nor how it is that they have become, in all quarters of the world, so deeply impressed on the mind of men; but it is worthy of remark that a belief constantly inculcated during the early years of life, whilst the brain is impressible, appears to acquire almost the nature of an instinct; and the very essence of an instinct is that it is followed independently of reason. (1871, p. 126)

Thus, according to Darwin's theory of the genesis of different moral judgments, each moral judgment arises through conscious or unconscious agreements or pacts made by the members of a population in order to privilege certain kinds of behavior (which are praised) above others (which are condemned).

Darwin's theory of the genesis of different moral judgments is compatible with his theory of the genesis of the enabling conditions for the moral capacity. Nevertheless, we must note that there might be many (virtually an infinite number of) different theories of the genesis of moral judgments, each compatible with Darwin's theory of the root of the moral capacity. For example, Nietzsche's theory of the genesis of different moral judgments is compatible, and so is Kitcher's (Chapter 9 in this volume). But Nietzsche's does not involve any explicit connection with biology, whereas Kitcher's is grounded on biological considerations.

Thus, if one is convinced by Darwin's argument for the roots of the moral capacity expounded in the previous section, but is unsatisfied with Darwin's

---

[7] It is interesting to note that usually anyone who tries to root moral judgments on animal social instinctual behavior neglects both the negative part of their "sociality" and the limits of such foundational inference. Darwin, instead, was totally aware of this fact: "It may be well first to premise that I do not wish to maintain that any strictly social animal, if its intellectual faculties were to become as active and as highly developed as in man, would acquire exactly the same moral sense as ours. . . . If, for instance, to take an extreme case, men were reared under precisely the same conditions as hive-bees, there can hardly be a doubt that our unmarried females would, like the worker-bees, think it a sacred duty to kill their brothers, and mothers would strive to kill their fertile daughters; and no one would think of interfering" (Darwin 1871, p. 102). On the "unsociality" of social insects, cf. Whitfield 2002.

explanation of the genesis of different moral judgments, one can maintain the former, while looking for a satisfying alternative to the latter. At this point it should be clear that there is an underdetermination of the theories of the genesis of different moral judgments by the theory of the biological root of moral capacity.

Darwin considered *The Origin* as "one long argument" in favor of evolutionary theory as a whole (1859, p. 435). Similarly, we could consider chapter 7 of *The Origin* and *The Descent* respectively as the proem and as a long argument in support of the idea that evolutionary theory implies a particular theory of the genesis of the enabling conditions for the human moral capacity. Moreover, *The Descent* may also be seen as containing a long argument in support of a particular theory of the genesis of different moral judgments.

This means that if we accept Darwin's evolutionary theory, we are forced to accept his theory of the biological roots of moral capacity, but we can reject his theory of the genesis of different moral judgments. Moreover, all of this allows us to state that:

1. Behavior is not intrinsically moral or immoral; morality and immorality are judgment-dependent properties.
2. We can both formulate and apply moral judgments and behave accordingly only because we are animals that have reached a suitable cerebral-mental evolutionary stage (we possess the enabling conditions).
3. The moral capacity is an evolutionary outcome that occurred in the phylogenesis of *Homo sapiens*, but it could occur also in the phylogenesis of other living beings.

The Consequences of a Darwinian Approach to the Moral Capacity

Following Darwin's indications, we have reached the conclusion that man's moral capacity must be seen as an evolutionary outcome and that there are no intrinsically moral behaviors but only moral judgments on behaviors. The latter statement means that there are behaviors that we consider moral because our moral judgments on them are positive (*proper moral behaviors*), and behaviors that we consider immoral because our moral judgments on them are negative (*improper moral behaviors*). We should be well aware that this implies (as Montaigne, Nietzsche and many other philosophers taught us, and as Villiers de L'Isle-Adam described so well) that what is a proper behavior for an individual or for a community might not be such for another individual or for another community. That is, although an instinctual, or noninstinctual, behavior may be independent of a particular culture and a particular community, this is not true of moral judgments.

Certainly, some could accuse me of moral relativism. Even if this is not the place to debate such a complex question, I wish to mention three points against this possible accusation.

First, accepting Darwin's evolutionary theory means accepting both Darwin's solution to the question concerning the biological roots of the moral capacity and the consequent separation between behavior and moral judgments on behavior. Therefore, if someone rejects the consequences, by *modus tollens* he should also reject Darwin's theory of evolution.

Second, my discussion does not imply that one must accept Darwin's theory on the genesis of different moral judgments. As already stated, there could be many theories of that sort, compatible with Darwin's theory of evolution, applied to the moral capacity.

Third, there are many kinds of relativism. Let us consider three of them:

1. Epistemological relativism, that is, the view that reason cannot find any *fundamenta inconcussa* for ethics
2. Phenomenological relativism (à la Montaigne), concerning the phenomenological awareness that different individuals and different communities might deliver different moral judgments on the same behavior
3. (Phylo)genetic relativism (à la Nietzsche, but also à la Darwin), regarding the idea that each moral judgment has a long history, beginning with an agreement, a convention, a habit, or something like that, among men

These kinds of relativism, however, must not be confused with *existential relativism* (i.e., the view of those who are indifferent to any particular moral hierarchy), or with the *relativism as to the consequences* (i.e., the idea that the consequences produced by certain kinds of behavior have no relevance with reference to our moral judgments). It should be clear that I am not suggesting a defense of existential relativism or of relativism as to the consequences. An epistemological relativist, a phenomenological relativist, or a (phylo)genetic relativist is not at all committed to the other two forms of relativism. For example, I might be an epistemological relativist, a phenomenological relativist, and a (phylo)genetic relativist (as I actually am). Nevertheless, like almost everyone (I suspect), I have a particular hierarchy of moral values and I take into great consideration the consequences of my (and other people's) actions; that is, I am neither an existential relativist, nor a relativist as to the consequences.

Let us move to another aspect. The evolution of what we call genotype can arrive at a stage in which the correlated physical phenotype – in particular, the cerebral-mental phenotype – allows us to make moral judgments on the instinctual and noninstinctual behavioral possible phenotype. This means that, from a certain stage of the history of evolution, there have been living beings with traits allowing them the moral capacity; that is, they had and have a suitable structure of the central and peripheral nervous system, a suitable structure of the nervous cells, a suitable genetic expression in the nervous cells, and a suitable regulation of that genetic expression.

This leads to two relevant remarks. First, in order to understand the correlations between biology and ethics, it seems that we should not be too concerned about the putative causal correlations between genes and behavior in general (Rosenberg, Chapter 10 in this volume). Instead, we should study the genetic and neuronal structures and processes that make both the formulation and the application of moral judgments and the coherent behavior possible, that is, that make the moral capacity possible (i.e., the enabling conditions). But there is also a second remark, and it concerns the weakness of moral will. An agent might judge a given behavior as morally improper and still fail to have the capacity for repressing it and, if necessary, for substituting it with a behavior that is judged morally proper. Are we sure that such an incapacity is always totally independent of biological aspects? Note that if there are genetic and neuronal enabling conditions for the moral capacity, then an individual whose genetic and neuronal enabling conditions are not "normal" could not have, or could have, a defective, "normal" moral capacity, where "normal" must be considered with great care. This should be neither forgotten nor neglected only on the basis of an abstract idea of political correctness but should be scientifically investigated and then experimentally tested.[8]

CONCLUSION

By distinguishing *behavior, moral judgment on behavior*, and *moral capacity*, I have first of all attempted to trace what a (neo-)Darwinian approach to the genesis of the moral capacity should be, by recalling Darwin's own statements. In doing this, I have contended that a theory of the roots of the moral capacity can be developed independently of any particular theory of the genesis of moral judgments. I have thus reached the conclusion that if one wants to be a (neo-)Darwinian, one necessarily should accept that the human moral capacity is an evolutionary outcome. But this has nothing to do with a possible biological reduction of moral judgments and systems.

---

[8] On scientific results connected with these possibilities, cf. Boniolo and Vezzoni, Chapter 5 in this volume; Pani 2000; Canali et al., Chapter 6 in this volume.

The foregoing considerations strongly suggest, however, that moral judgments are not grounded on something mysterious or numinous, but in human
affairs and human interrelations. Therefore, to paraphrase Nietzsche, nothing
is *Menschliches, Allzumenschliches* as the formulation and the application
of moral judgments, and nothing is *Menschliches, Allzumenschliches* as the
capacity for formulating and applying them. Yet, only the latter requires analysis in biological terms.

REFERENCES

Darwin, C. 1859. *The Origins of Species*. New York: Gramercy Books, 1976.
Darwin, C. 1871. *The Descent of Man*. Amherst, N.Y.: Prometheus Books, 1998.
Futuyma, D. J. 1979. *Evolutionary Biology*. Sunderland, Mass.: Sinauer Associates.
Montaigne, M. 1572–1592. *The Essays*. Trans. E. J. Trechmann. Oxford: Oxford Uni-
versity Press, 1927.
Nietzsche, F. 1887. *On the Genealogy of Morals*. In *On the Genealogy of Morals and
Ecce Homo*. New York: Random House, 1967.
Pani, L. 2000. Is There an Evolutionary Mismatch between the Normal Physiology of
the Human Dopaminergic System and Current Environmental Conditions in Indus-
trialised Countries? *Molecular Psychiatry* 5: 467–475.
Plato. 1997. *Phaedrus*. In Plato, *Complete works*, ed. J. M. Cooper. Indianapolis, Ind.:
Hackett.
Ridley, M. 2004. *Evolution*. Oxford: Blackwell.
Thomasma, D. C., and Weisstub, D. N. (eds.). 2004. *The Variables of Moral Capacity*.
Dordrecht: Kluwer.
Villiers de L'Isle-Adam, J.-M.-M.-P.-A. 1874. Les damoiselles de Bienfilâtre. In *Contes
crues* (1883). Paris: Gallimard, 1986.
Whitfield, J. 2002. The Police State. *Nature* 416: 782–784.

II

Methodological Issues Concerning Evolutionary Accounts of Ethics

3

Are Human Beings Part of the Rest of Nature?

CHRISTOPHER LANG, ELLIOTT SOBER, AND KAREN STRIER

The issue we want to address is not whether human beings should be understood naturalistically or *super*naturalistically. Rather, our question concerns the kinds of naturalistic explanations that are needed to account for the features that human beings exhibit. If a factor C helps explain some feature E of nonhuman organisms, should we infer that C also helps explain E when E is present in human beings? The choice that interests us is between *unified* and *disunified* explanations. Do human beings fall into patterns exhibited by the rest of nature, or are we the result of fundamentally different causal processes?[1]

Although evolutionary theory is often seen as the vehicle for understanding human beings as part of the natural order, it would be wrong to assume that evolutionary explanations are automatically unified. An evolutionary explanation for why two species have a feature need not claim that they have that feature for the same reason. Fir trees are green and so are iguanas, and there is an evolutionary explanation for each of these outcomes; however, iguanas and fir trees are green for very different evolutionary reasons. In fact, within an evolutionary framework there are four possible patterns of explanation, not just two; these can be described by beginning with the three options depicted in Figure 3.1.

In case 1, the two species (S_1 and S_2) are similar because they inherited their shared feature from a common ancestor (A); the similarity is a *homology*. In both cases 2 and 3 the two descendant species obtained feature E

---

[1] Thus the question we are considering is the mirror image of the problem of self-to-other inference that constitutes the traditional philosophical problem of other minds. See Sober 2000 for discussion.

We are grateful to James Crow, Branden Fitelson, Elisabeth Lloyd, Larry Shapiro, and David Sloan Wilson. Part of this essay is reprinted with the kind permission of *Biology and Philosophy*.

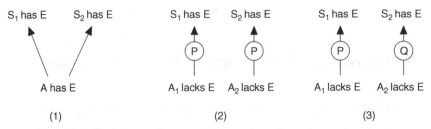

Figure 3.1. Evolutionary framework: comparison of homology and analogy.

by independent evolution; the similarity is an *analogy*.[2] Within this category of analogous similarities, one can distinguish functionally similar analogies from functionally dissimilar analogies (Sober 1993). Even though birds and bats evolved their wings independently, it still may be true that the trait evolved for the same reason (P) in the two lineages – in both instances, wings evolved because they facilitated flight (case 2). The green coloration of fir trees and iguanas is different. Not only is the similarity not homologous; in addition, the reason the color evolved in the lineage leading to fir trees differs from the reason it evolved in the line leading to iguanas (P ≠ Q); this is case 3.[3]

In Figure 3.1, inheritance from a common ancestor (case 1) is represented as a possibility on a par with the two types of analogy depicted in cases 2 and 3, but in fact the category of homology needs to be subdivided. If two descendant species have trait E because their most recent common ancestor had E, it is a further question as to why the trait was maintained in the two lineages. It is possible that the trait was retained in the two lineages for the same reason (P), or for different reasons (P ≠ Q) Thus, case 1 in Figure 3.1 needs to be separated into the two scenarios depicted in Figure 3.2.

Inheritance from a common ancestor is often thought of as a nonselective explanation of a trait's presence; however, the fact of the matter is that a descendant can exhibit the trait possessed by its ancestor for selective as well as for nonselective reasons (Orzack and Sober 2001). Stabilizing selection can cause stasis. But what does it mean for a descendant to have a trait, not because of stabilizing selection, but simply because its ancestor had the trait? We take this to mean that the trait was retained because there wasn't sufficient time for the descendant to evolve away from the ancestral condition. If so, it is appropriate to talk of *ancestral influence* or *phylogenetic inertia* (Harvey

---

[2] Although cases 2 and 3 in Figure 3.1 do not depict a common ancestor, we assume that one exists.
[3] It is worth noting that the pattern in case 3 subdivides into two possibilities; there may be a partial overlap between the reasons the trait evolved in the two lineages, or the reasons may be entirely *disjoint*.

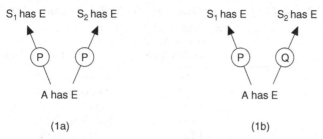

Figure 3.2. Evolutionary framework: homology.

and Pagel 1991; Orzack and Sober 2001). Although selection and inertia are different possible causes of stasis, they are compatible; both can contribute to a trait's retention, as Table 3.1 indicates. Ancestral influence occurs when a lineage's initial condition affects its subsequent state; selection, on the other hand, is a process that occurs during the duration of the lineage. Both the lineage's initial condition and the processes that then set to work can affect the character states of descendants.

We so far have described how we understand the question of whether human beings are part of the rest of nature by considering how one should explain a *similarity* that unites human beings and one or more nonhuman species. Explanations in cases 1a and 2 are unified; explanations in cases 1b and 3 are disunified. However, the choice between unified and disunified explanations also arises when one wants to explain why the species under consideration exhibit *different* trait values. Rather than develop this point abstractly, we can explain it in terms of an example.

In modern industrial societies, women on average live longer than men. One might suspect that this is a recent phenomenon, a result of improved medical care that reduces the risk of dying in childbirth. In fact, the data available suggest otherwise. In eighteenth-century Sweden, for example, women lived longer than men, and this inequality continued right up to the present, despite a steady improvement in the longevities of both sexes. The same is true of the

Table 3.1. *Why Was Trait E, Which Was Found in the Ancestor, Retained in the Descendant?*

	Natural Selection	
	For Trait E	**Against Trait E**
Phylogenetic inertia		
Little time	Both selection and inertia	Inertia only
Lots of time	Selection only	Neither

Table 3.2. *Survival Ratios and Male Care of Offspring in Anthropoid Primates*

Primate	Female-Male Survival Ratio	Male Care of Offspring
Chimpanzees	1.418	Rare or negligible
Spider monkey	1.272	Rare or negligible
Orangutan	1.203	None
Gibbon	1.199	Pair-living, but little direct role
Gorilla	1.125	Protects, plays with offspring
Human (Sweden 1780–1991)	1.052–1.082	Supports economically, some care
Goeldi's monkey	0.974	Both parents carry offspring
Siamang	0.915	Carries offspring in second year
Owl monkey	0.869	Carries infant from birth
Titi monkey	0.828	Carries infant from birth

Source: Reproduced with permission from J. Allman, A. Rosin, R. Kumar, and A. Hasenstaub, "Parenting and Survival in Anthropoid Primates – Caretakers Live Longer," *Proc. Natl. Acad. Sci. USA* 95: 6866–6869, Copyright (1998) National Academy of Sciences, U.S.A.

Ache, a hunter-gatherer group now living in Paraguay. Indeed, in twentieth-century societies around the world, one almost always observes that women live longer than men. Is this fact about human beings to be explained in terms of some constellation of causes that is unique to our species? Or is the pattern of longevity in human beings due to factors that apply to a more inclusive set of organisms?

Allman et al. (1998) cite the facts just described and seek to explain them in terms of a general hypothesis about anthropoid primates – when one sex provides more parental care than the other, selection favors reduced mortality in the sex that makes the larger contribution. They hypothesize that selection will generate a quantitative relationship – the greater the imbalance in parental care, the more skewed the longevity should be in favor of the sex that provides more parental care. Although they do not spell out their reasoning in much detail, their idea is presumably that the sex that provides more parental care would incur a greater fitness cost by accepting an increased risk of mortality; this leads the sex that provides more parental care to be more risk-averse. In support of their hypothesis, the authors present the data in Table 3.2.

Allman et al. wanted to test the hypothesis that disparity in parental care causes disparity in longevity – the latter is an adaptive response to the former. They cite as confirmation the fact that the two variables are associated in the data, and we do not disagree. However, it is important to understand this evidential claim in the right way. Allman et al. tested their causal hypothesis against a null hypothesis, one that says that the two variables are causally unrelated. The former hypothesis predicts an association in the data, whereas the latter predicts no association.[4] This methodology is fine as far as it goes, but it has its limitations. The data do not favor the causal hypothesis that Allman et al. formulate over its converse – that differences in longevity caused differences in parental care.[5] Nor do the data rule out the hypothesis that the two variables are effects of a common cause.

The first thing to notice about this pattern of argument is that the exact survival ratio exhibited by human beings differs from that found in other species. Allman et al. are arguing that human beings are "part of the rest of nature," but this does not mean that the human characteristic they wish to explain must be identical with the characteristics found in other species. Rather, the study defends a unified account of the human trait value by showing how the human value falls within a larger pattern of variation. The point is that *we are not outliers*. We may be unique in our trait value (just as other species are in theirs), but the suggestion is that we are not unique with respect to the causal processes generating that trait value.

Notice also that Allman et al. do not attempt to explain the pattern of variation that exists within our species, or, for that matter, the variation found in other species. The trait value for a species is the species *average*. It is perfectly consistent with their analysis that the ratio of female-to-male longevities should fail to be positively related to the ratio of female-to-male parental care as one looks across populations within our species. Would this show that human beings are not "part of nature"? Here we must recognize the limited usefulness of this way of posing the question. Not only must we relativize our question about the place of human beings in nature to a specific trait (in this instance, the fact that women live longer than men). In addition, we need to specify the pattern of variation that we wish to consider. It is entirely possible that the human *average* fits in with data about the average values found in other species, even if human *variation around that average* is generated by causal processes that differ fundamentally from the factors that generate variation within or among other species.

---

[4] The argument of Allman et al. exhibits a pattern of argument that is entirely standard in evolutionary biology. Although hypotheses about natural selection purport to describe processes at work within lineages, the data sets used to test those hypotheses usually describe the character states of tip species. Why should the latter be able to confirm or disconfirm the former? For discussion, see Sober and Orzack 2003.

[5] It sometimes is possible to discriminate between the hypothesis that E is an adaptive response to C and the hypothesis that C is an adaptive response to E by seeing which trait evolved first. This procedure requires one to reconstruct the character states of ancestors in a phylogenetic tree. Cladistic parsimony is the method usually used to do this; see Sober 2002 for discussion.

Let us consider what it would mean if cross-cultural variation in the longevity ratio were positively associated with the ratio of contributions to parental care. This could be true even if the human average were an outlier in the context of cross-species data. Furthermore, the existence of human plasticity does not automatically place us "outside the rest of nature." Even if, contrary to fact, ours were the only species that exhibits within-species variation in these features, it still could be true that we are part of the larger picture. What would be unique about us is our *plasticity;* but the factors that explain within-species variation in our case could still coincide with the factors that explain the pattern of between-species variation.[6]

The logical independence of these two levels of analysis – within–species and between-species – is depicted in Figure 3.3. We can ask how the human average relates to average values found in other species. And we can ask how variation within our species relates to those average values. The latter may seem like an "apples and oranges" question, because we are comparing within-species variation with between-species variation. Nonetheless, the question makes sense and has its point. In order to keep things simple, we have omitted a third question from Figure 3.3, one that is logically independent of the first two – how does within-species variation in our species relate to within-species variation in other species? The two questions that are described in Figure 3.3 generate four possibilities, which differ with respect to whether and how the human condition is unified with the situation found in the rest of nature. We have included in each cell of the figure a hypothetical data set that would support the relevant interpretation.[7]

Several features of this framework merit comment. First, it is important to see that it is specific models about the relationship of specific dependent and independent variables that get tested; the bare claim that there exists a unified

[6] Within-species variation could be due to genetic variation, environmental variation, or both; all three possibilities are consistent with the adaptive hypothesis (Sober 1993). There need be no commitment to "genetic determinism."

[7] The epistemology of choosing between unified and disunified explanations is interesting. Even if human beings are not outliers, why couldn't it be true that human trait values are the product of fundamentally unique causal processes? Conventional frequentist statistics treats the unified explanation as a null hypothesis, one that asserts that there is no difference between the human and nonhuman causal situations; testing takes the form of asking whether the data permit one to reject this null hypothesis. According to this approach, one should embrace the disunified explanation only if the data force one to do so. Frequentist statistics thus assigns a privileged status to unified explanations. For discussion of Bayesian approaches to this problem, see Forster and Sober 1994, which also locates the problem within the framework of Akaike's criterion for model selection. Some such statistical framework is needed to define what it means for a species to be an "outlier" – how much distance between the human trait value and the regression line for other species must there be for this to be true?

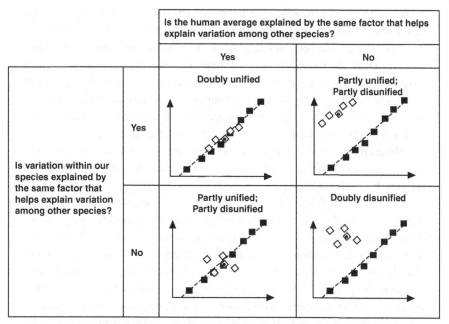

Figure 3.3. Explanations of patterns of variation. Each white diamond represents the average value for a human population. The white diamond with a dot in it represents the human average. Each black square represents the average value for a nonhuman species. The line is the best-fitting regression line for the nonhuman species.

(or a disunified) explanation of some effect (e.g., why women live longer than men) does not make testable predictions. A related point is that it is possible for human beings to fall into the pattern represented in one cell of Figure 3.3 when X and Y are the variables used, whereas the pattern changes to that depicted in another cell when a new independent variable Z is used instead of Y. Finally, we note that specific models make probabilistic (not deductive) predictions about data; this means that it is perfectly possible that a given model is true and yet the data one observes fail to conform to the patterns associated in Figure 3.3 with that model.

What would it mean if human beings were outliers twice over (as depicted in the lower right cell of Figure 3.3) – suppose our average trait value does not conform to the pattern displayed by other species, and suppose that variation within our species exhibits a different pattern from that found among other species? This may be due to the fact that we human beings are influenced by nonbiological, cultural forces that are unique to us. Or it may be that biological causes distinct from those acting on other species are at work. Our deviating from patterns found in the rest of nature does not decide this question.

We began by discussing the question of whether human beings are part of the rest of nature by examining possible explanations of the *similarities* that may unite human beings and this or that nonhuman species. We then explained how the same question can arise in explaining why human beings have trait values that *differ* from those found in other species. The key here is not whether human beings have the same traits or different ones, but how the distribution of characters among species is explained. Similarities can arise from different causes and differences can arise from the same causes. The issue concerns causation, not whether we are similar to or different from other organisms. This question is not settled by the fact that human beings are genealogically related to other species.

We want to emphasize the role played by the concept of variation in our analysis of how the place of human beings in nature should be assessed. To decide whether human beings are part of the causal pattern found in the rest of nature, or deviate from it, one must be able to identify what that wider causal pattern is. For this to be possible, there must be variation in one's data. If a quantitative variable X causes a quantitative variable Y, then changes in X values should be associated with changes in Y values (once one controls for other contributing causes).[8] If smoking causes cancer, then people who smoke more should get cancer more frequently than people who smoke less, where comparisons are carried out among individuals who are otherwise the same with respect to other factors that influence cancer. The causal proposition would not be tested by a data set in which everyone smoked to the same degree. By the same token, the causal hypothesis advanced by Allman et al. – that disparities in parental care cause disparities in longevity – would not be tested by a data set in which the ratio of female-to-male investment is the same across the species considered.

The simple fact that testing causal hypotheses requires a comparison of different types of cases – of situations in which dosages of the putative causal factor are different – helps explain part of what Gould and Lewontin (1979) were getting at when they criticized the invention of "just-so stories" in evolutionary biology. When one's observation is the simple fact that species S has trait T, the data are too impoverished to provide a proper test of a causal explanation. It isn't that adaptive hypotheses are untestable, but rather that it takes a certain kind of data set to put them to the test. If the dichotomous trait T is universal within species S, then one needs a data set in which some species

<hr>

[8] Allman et al. do not investigate whether the similarities they observe were due partly to phylogenetic inertia rather than adaptation. Testing an adaptive hypothesis requires that one control for this possibility. See Felsenstein 1985 and Orzack and Sober 2001 for discussion.

have trait T while others do not. In one sense, the explanation of impoverished data is easy – it is easy enough to invent a story that fits the data. In another sense, however, the explanation of impoverished data is impossible – the data do not permit adaptive hypotheses – to be tested properly. This, we suggest, is what it means for adaptive storytelling to be "too easy."[9]

Adaptationism, as Gould and Lewontin understand that -ism, contrasts with evolutionary pluralism. Gould and Lewontin say they agree with Darwin that natural selection has been the *most important cause* of evolutionary change. What they disagree with is the monistic idea that natural selection has been the *only important cause*. We interpret this adaptationist hypothesis to predict that organisms should have locally optimal traits; they should exhibit traits that are fitter than any of the available alternatives (Orzack and Sober 1994; Sober 1993). Given this contrast between adaptationism and pluralism, it is important to recognize that the hypothesis that Allman et al. were testing is *not* an instance of adaptationism. They were not claiming that the disparity in mortality rates between the sexes is optimal; indeed, their article does not even specify what the optimal disparity would be. The argument of Allman et al. was merely to show that male-female differences in longevity were influenced by natural selection. Because pluralists just as much as adaptationists are committed to the importance of natural selection, both need to avoid telling "just-so stories" about that process; in this sense, adaptive hypotheses are not the exclusive property (and problem) of adaptationists. Data sets that exhibit variation are a useful prophylactic device; they make it harder to invent adaptive scenarios.

In addition to throwing light on the general problem of testing adaptive hypotheses, the methodology we are suggesting also elucidates a special problem that arises in connection with human evolution. To test causal hypotheses about the place of human beings in nature, the human traits under study must be commensurable with the traits found in other species. As we saw in connection with the data used by Allman et al., it is not essential that human beings and other organisms have exactly the same trait values. Rather, the point is

---

[9] Gould and Lewontin also say that if one adaptive hypothesis fails, another can be invented in its place, and that this possibility constitutes a flaw in adaptationism. The first thing to notice about this claim is that it envisions adaptive hypotheses' failing; this presupposes that a data set is being consulted that is not impoverished – it succeeds in putting the hypothesis to the test. We also note that the possibility that Gould and Lewontin describe in connection with adaptationism also is possible for the evolutionary pluralism they advocate – if one pluralistic model fails, another can be invented in its place. Because this is a feature of all research programs, it does not constitute a reason for rejecting any particular research program, though it does raise the question of when a research program ceases to be worth pursuing (Sober 1993).

that their study used quantitative variables – the female-to-male survival ratio and the female-to-male ratio of parental care – that subsume human beings and other species alike. The same point would apply if the traits considered were dichotomous. If human beings and other species can be said to have or lack trait C, and the same is true of trait E, then a data set can be obtained that allows one to evaluate whether C and E are associated. But suppose human beings are the *only* species that exhibits trait E. If so, there is an easy recipe for finding causal hypotheses that fit the data – merely find a trait C that also is unique to human beings. The result is that C and E will be perfectly associated in one's data. The trouble is that the data will not help one pry apart different causal hypotheses that focus on different uniquely human features. The hypothesis "C_1 causes E" will fit the data, but so will "C_2 causes E," "C_3 causes E," and so on. Here we find a second context in which the invention of adaptive hypotheses is too easy. If there is no variation in one's data, the data are useless. But if the variation is such that all species are the same, save one, the data are next to useless.

We think there is an important lesson here for the research program known as evolutionary psychology. Tooby and Cosmides (1990) argue that evolutionary theory predicts that the complex adaptive features found in our species (or, indeed, in any species) will be species-typical universals. There is a great deal of room to doubt whether evolutionary theory provides principled reason to expect there to be no within-species adaptive variation (Wilson 1994). Furthermore, the argument made by Tooby and Cosmetics has not stopped other evolutionary psychologists from trying to find adaptive explanations of behavioral differences between the sexes (Buss 1994; Daly and Wilson 1988), and this, of course, is an instance of within-species variation. The point we want to make here, however, is that if the trait of interest is universal in our species, then the only way to explain its presence is to adopt a comparative perspective. Otherwise, one is working with a single data point – an impoverished data set if ever there was one. The problem is that the features that often interest evolutionary psychologists are uniquely human – for example, the human language faculty, or the cognitive capacity to analyze what kinds of observations would falsify a conditional statement. This leaves it open that evolutionary psychologists can attempt to embed their description of human beings within a wider view of the traits found in other species. For example, perhaps there are features of human language that can be related to features of communication systems used in other species.[10] The other way forward for evolutionary psychology is to focus on traits with respect to which human

---

[10] This type of analysis was developed by Pinker and Bloom 1990 and by Pinker 1994.

beings vary. What is a dead end, in our view, is the attempt to explain human universals that are unique to our species without relating those features to trait values found in other species.

Although our protocol for testing hypotheses concerning the place of human beings in nature has focused on adaptive hypotheses, our proposed methodology is not limited to hypotheses that make claims about natural selection. It isn't just adaptive hypotheses that can be elaborated as "just-so stories." Hypotheses that postulate nonadaptive processes also require data sets that are not impoverished. For example, consider the hypothesis that the human language faculty did not evolve because it facilitates communication but was merely a by-product of the evolution of a big brain, which evolved for other adaptive reasons.[11] We submit that this proposition is untestable if one looks just at human beings. And if one considers a range of species within which the language faculty and a big brain are both unique to our species, the by-product hypothesis is testable, but now a new difficulty arises, one that we described previously. The invention of nonadaptive hypotheses also can be "too easy."

We have emphasized that the data sets used to test adaptive hypotheses must contain variation. Is there an alternative methodology, one in which a single observation of a trait that is universal within a species suffices? We are skeptical. Granted, if an adaptive hypothesis specifies the optimal trait value and asserts that organisms have attained their optima, it is possible to make a single observation and determine whether the model's prediction is correct. As we have emphasized, however, adaptive hypotheses are very often not of this form. What these weaker hypotheses assert is that species have evolved in the direction of more-optimal trait values. These hypotheses predict that lineages have *changed* in certain ways. Hypotheses about the direction of change cannot be tested by a single snapshot. Even when a species is "close" to the trait value that an optimality model says is optimal, the question of whether the lineage leading to that species has evolved toward that trait value or away from it remains (Sober and Orzack 2003).

Our demand for comparative data may seem gratuitous when it seems obvious that a species' trait value is an adaptive response to some problem that the species faces. It may seem obvious that the polar bear's thick fur evolved

---

[11] Thus Gould (1991: 62): "the traits that Chomsky (1986) attributes to language – universality of the generative grammar, lack of ontogeny, ... highly peculiar and decidedly nonoptimal structure, formal analogy to other attributes, including our unique numerical faculty with its concept of discrete infinity – fit far more easily with an exaptive, rather than an adaptive, explanation. The brain, in becoming large for whatever adaptive reasons, acquired a plethora of cooptable features, Why shouldn't the capacity for language be among them?"

as an adaptive response to cold weather. There seems to be an obvious "fit" between the warm coat and the icy temperature. Why do we have to look at bears that live in warmer climates to see if they have thinner coats? The polar bear's fur and the temperature of the bear's environment resemble a key and the lock it opens. Shouldn't it be self-evident that the one was made as a solution to the problem posed by the other?[12] There is reason to be cautious here, however, because every biologist can recount examples in which intuitively "obvious" adaptive scenarios turned out to be disconfirmed by data. Without a correlation between fur thickness and ambient temperature, the hypothesis that polar bears have thick fur as an adaptive response to ambient temperature remains a mere plausible conjecture.[13]

In closing we want to describe a kind of problem that the methodology we have proposed does not solve. If an adaptive hypothesis predicts a correlation, it can be tested against a null hypothesis that predicts no correlation. And if a nonadaptive hypothesis predicts a correlation, it too can be tested against a null hypothesis that predicts no correlation. But how can adaptive and non-adaptive hypotheses be tested against each other? If they predict *different* correlations, the procedure is straightforward. But what if they predict *the same* correlation?[14] This is an interesting question, but it differs from the one we set out to address. Notice that the adaptive and the nonadaptive hypotheses we now are considering are both *unified* – both seek to explain human trait values by situating them in a causal framework that subsumes other species as well. Our concern in this chapter has been to discuss how unified and dis-unified hypotheses should be compared, not to make assessments within the category of unified explanations.

REFERENCES

Allman, J., Rosin, A., Kumar, R., and Hasenstaub, A. 1998. Parenting and Survival in Anthropoid Primates – Caretakers Live Longer. *Proc. Natl. Acad. Sci. USA* 95: 6866–6869.

Buss, D. 1994. *The Evolution of Desire: Strategies of Human Mating.* New York: Basic Books.

Chomsky, N. 1986. *Knowledge of Language: Its Nature, Origins, and Use.* New York: Praeger.

---

[12] Lewontin's (1982) skepticism of the idea that adaptations are "solutions" to preexisting "problems" by the environment is relevant here.

[13] In this connection, see Sober's (1997) discussion of how one might test Godfrey-Smith's (1996) hypothesis that mental representation is an adaptation for coping with environmental complexity.

[14] The study of Allman et al. provides an example. Their hypothesis is that disparities in parental care cause disparities in longevity. How would one test this against the claim that the two traits are pleiotropic consequences of the same set of genes?

Daly, M., and Wilson, M. 1988. *Homicide*. New York: Aldine de Gruyter.

Felsenstein, J. 1985. Phylogenies and the Comparative Method. *American Naturalist* 125: 1–15.

Forster, M., and Sober, E. 1994. How to Tell When Simpler, More Unified, or Less *Ad Hoc* Theories Will Provide More Accurate Predictions. *British Journal for the Philosophy of Science* 45: 1–36.

Godfrey-Smith, P. 1996. *Complexity and the Function of Mind in Nature*. Cambridge: Cambridge University Press.

Gould, S. 1991. Exaptation – a Crucial Tool for an Evolutionary Psychology. *Journal of Social Issues* 47: 43–65.

Gould, S., and Lewontin, R. 1979. The Spandrels of San Marco and the Panglossian Paradigm – a Critique of the Adaptationist Programme. *Proceedings of the Royal Society of London B* 205: 581–598.

Harvey, P., and Pagel, M. 1991. *The Comparative Method in Evolutionary Biology*. Oxford: Oxford University Press.

Lewontin, R. C. 1982. Organism and Environment. In *Learning, Development, Culture*, ed. H. Plotkin, 151–170. New York: Wiley.

Orzack, S., and Sober, E. 1994. Optimality Models and the Test of Adaptationism. *American Naturalist* 143: 361–380.

Orzack, S., and Sober, E. 2001. Adaptation, Phylogenetic Inertia, and the Method of Controlled Comparisons. In *Adaptationism and Optimality*, ed. S. Orzack and E. Sober, 45–63. Cambridge: Cambridge University Press.

Pinker, S. 1994. *The Language Instinct*. New York: William Morrow.

Pinker, S., and Bloom, P. 1990. Natural Language and Natural Selection. *Behavior and Brain Sciences* 13: 764–765.

Sober, E. 1993. *Philosophy of Biology*. Boulder: Westview Press.

Sober, E. 1997. Is the Mind an Adaptation for Coping with Environmental Complexity? A Review of Peter Godfrey-Smith's *Complexity and the Function of Mind in Nature*. *Biology and Philosophy* 12: 539–550.

Sober, E. 2000. Evolution and the Problem of Other Minds. *Journal of Philosophy* 97: 365–386.

Sober, E. 2002. Reconstructing Ancestral Character States – a Likelihood Perspective or Cladistic Parsimony. *The Monist* 85: 156–176.

Sober, E., and Orzack, S. 2003. Common Ancestry and Natural Selection. *British Journal for the Philosophy of Science* 54: 423–437.

Tooby, J., and Cosmides, L. 1990. On the Universality of Human Nature and the Uniqueness of the Individual – the Role of Genetics and Adaptation. *Journal of Personality* 58: 17–67.

Wilson, D. S. 1994. Adaptive Genetic Variation and Human Evolutionary Psychology. *Ethology and Sociobiology* 15: 219–235.

4

The Nature of Resemblance

Homologues in the Nervous System and Behavioral Correspondence

ALDO FASOLO

It would be very difficult for any one with even much more knowledge than I possess, to determine how far animals exhibit any traces of . . . high mental powers. This difficulty arises from the impossibility of judging what passes through the mind of an animal; and again, the fact that writers differ to a great extent in the meaning which they attribute to the above terms, causes a further difficulty.

Darwin 1871, p. 85

INTRODUCTION

Morality is related variously to behavioral tendencies and cognitive capacities. In this chapter, I do not discuss what the forms and modes of these relations can be. Instead, I focus on the topic of the homologues in the nervous system and then consider whether and to what extent the capacities that are relevant for morality can be explained through biological comparisons. Nevertheless, the quest for the biological basis of behavior and cognition must rely on a sound comparative approach, satisfying Theodor Bullock's three Rs requirement (roots, rules, relevance; cf. Bullock 1983) for the study of cognition and highly integrated behavior. The satisfaction of this requirement, indeed, seems to be the only way to discern scientifically plausible biological hypotheses from merely intriguing scientifically sounding metaphors.

In what follows, I argue that any attempt to explain (human) ethics through considerations concerning animal behavior is hopeless, unless we manage to individuate strict homological correspondences among all the feasible generic analogies between human behavior and animal behavior.

This conclusion is reached through an argument that involves five theses. First, the possibility to develop a comparative approach depends on the concept of homology, which is arguably central in comparative biology (Wake 1994). Second, homological recognition and related methods are not only essential for morphological analysis and taxonomy but also for an understanding of the evolution of brain and cognition. Third, novel, interesting hints on how to interpret similarities can be obtained from evolutionary and developmental (Evo-Devo) studies. Fourth, the concept of modularity has to be redefined, in the light of developmental neurobiology. Furthermore, it is plausible that such modularity does not fit into the selectionistic stance of evolutionary psychology. Fifth, a comparative analysis and a critical survey of current theories of brain evolution lead to the conclusion that, besides adaptation, other processes help to envisage the multiple forces shaping complex behavior, such as exaptation or coevolution.

The upshot of my conclusion is that an explanation of the emergence of the human moral capacities throughout evolution cannot merely rely on simplistic analogies (sometimes erroneously called homologies) between *Homo sapiens* and other animal species. The interesting point would be the possibility to identify the characters, if any, that show continuity and can be challenged by a homological analysis.

Character Comparison: Concepts and Application

Homology and its related concept of analogy have been central in comparative morphology (Minelli 2003), because homologous structures are integrated as fundamental parts of a given body plan; they become "attractors of morphological design," a sort of backbone to which further elements of the body design may be added (Müller and Newman 1999).

In addition, with the load of data and speculations coming from genomics and bioinformatics, the concept of homology was enormously extended, because it began to be employed for comparing molecular structures. This field is dominated by an "extrapolationist paradigm" that is guided by the assumption that homologous structures can have similar functions, and it introduced the somewhat misleading concept of "functional homology." For instance, someone writes that "finding homologous genes, or homologs, shared between two species allows one to make inferences about the function of the gene in one organism by extension from the other organism. This kind of extrapolation of function to homologous sequences in another organism forms the bulk of functional annotation in large sequence data bases" (Striedter 2002) and has tremendous practical applications. Nevertheless, the idea of grounding inferences on molecular similarities is a conceptual oversimplification, which may lead to fallacious forms of comparative reasoning. Indeed, Wray (1999) argued that evolutionary dissociations between homologous genes and homologous structures are a viable possibility.

This possibility leads to more complex structures of reasoning: "the ability to describe phylogenetic changes (i.e., any definable attribute of an organism in any character) is based on the pattern of variation observed among different taxa. Equally important, the elucidation of evolutionary mechanisms or processes is based on the kinds of character patterns that can be recognized. In both of these analyses it is critical to distinguish a character and its subsequent phylogenetic transformation (homologous characters) from other characters that may appear similar, but have different evolutionary histories (homoplasous characters) if errors in interpretation are to be minimized" (Northcutt 1984, p. 701). In the same review, Northcutt discussed the concept of character comparison in depth, focusing on character similarity and common ancestry. He finally adopted the following definitions, taken from Wiley (1981),

1. *Homology*: "A character of two or more *taxa* is homologous if this character is found in the common ancestor of these *taxa*, or, two characters (or a linear sequence of characters) are homologous if one is directly (or sequentially) derived from the other(s)."
2. *Homoplasy*: "A character found in two or more species is homoplasous (non-homologous) if the common ancestor of these species did not have the character in question, or if one character was not the precursor of the other."
3. *Convergence*: "[It] is the development of similar characters from different pre-existing characters."

These apparently clear-cut definitions, though, did not help much in solving a problem about homology that dates back to Darwin's times: homology has always been defined in one way and tested in another. In particular, definitions of homology have been based on common ancestry, but the criteria for homologue recognition have generally rested on phenetic similarity. In cladistic analysis, for example, features are considered homologous when they characterize monophyletic groups (Patterson 1982). In order to recognize shared derived characters, it is assumed that one must determine the direction of change or polarity (i.e., primitive versus derived condition) of the characters that are suspected to be homologous on the basis of phenetic similarity.

Three criteria are usually given: outgroup rule, ontogenetic character precedence, and geological character precedence. The first two criteria are possibly quite useful in comparative analysis, and they may need some explanation. The outgroup rule states that "given two characters that are homologues and found within a monophyletic group, the character that is also found in the sister group is the primitive (plesiomorphic) character, whereas the character found only within the monophyletic group is the derived (apomorphic) character" (Northcutt 1981). The criterion for ontogenetic character precedence (derived from von Baer's theorem) states that character polarity has to be determined on the basis of the comparison of developmental patterns, rather than the distribution of characters among adults in closely related taxa. Von Baer's theorem states that members of two or more closely related taxa will follow the same course of development to the stage of their divergence. Thus characters observed to be more general are assumed to be primitive, whereas those that are less general are assumed to be derived.

The puzzling point remains the underskin concept of phylogenetic continuity. To paraphrase the elegant essay by Striedter (1998), what does it exactly mean for homologous characters to exhibit phylogenetic continuity? As Bateson (1892, 1894) pointed out, evolutionary biologists frequently seem to think of phylogenetic transformations as if they were direct transformations between characters in their adult state. De Beer (1971) suggested that homologous characters might differ in their genetic basis, and then asked "what mechanism can it be that results in the production of homologous organs, the same 'patterns,' in spite of their not being controlled by the same genes?" Until that question is answered, de Beer concluded, homology must remain an "unsolved problem." Goodwin (1984) proposed a new "generative paradigm," according to which morphological homology is simply structural correspondence, as in pre-Darwinian times, and structural correspondences exist because the number of possible morphologies is severely limited by the generative mechanisms of development. Because these generative rules are assumed to be constant and universal, Goodwin considers evolutionary explanations of homology to be mere statements about history. A less extreme, but also less rigorous, solution to de Beer's problem was offered by van Valen (1982), who argued that homology may be "defined, in a quite general way, as correspondence caused by continuity of information."

The Evo-Devo Approach and Modularity

Recently, two rather contrasting views of homology emerged, "phylogenetic" versus "developmental" homology (Striedter 1998). According to Wagner (1994) and Butler and Saidel (2000) we can go further and recognize three kinds of homology, based on the questions that they are expected to answer:

1. Historical (phylogenetic) homology, dealing with character distribution among taxa
2. Biological homology, describing mechanisms of character evolution
3. Generative homology, defining processes of character development

In addition, Striedter (1998) stressed a novel interest in old concepts, like the *epigenetic landscape* of Waddington (1957), which may be useful for understanding the interplay between genes, epigenetic pathways, and the final phenotypes. The so-called *epigenetic homology* (which considers homologues as recurring attractors in epigenetic landscapes) has profound implications for comparative neurobiology and explains well, for instance, variations in monkey visual cortex after developmental perturbations, which is a classical case in neural plasticity and brain reorganization studies.

Within these debates, there is a growing agreement about the existence of developmental and evolutionary modules and about their importance to understand the evolution of morphological phenotypes (Wagner et al. 2005; West-Eberhard 2003). Modules are considered important for the evolution of complex organisms (Wagner and Altenberg 1996) and the identification of independent characters (Houle 2001; Kim and Kim 2001; Wagner 1996). Modules are altogether necessary for explaining heterochrony (Gould 1977).

Wagner et al. (2005) argued that the empirical basis for *developmental modules* is the observation that embryos can develop certain parts quite independently from the environment in which they are hosted. This is the case of limb buds, tooth germs, developmental fields, and clusters of interacting molecular reactions. On the other hand, *evolutionary modules* are characterized by a variational independence from each other, and by the integration among their parts, either in interspecific variation or in mutational variation (Wagner and Altenberg 1996). According to a preliminary definition by Wagner, an evolutionary module is a set of phenotypic features that are highly integrated by the pleiotropic effects of the underlying genes and are relatively isolated from other sets by a paucity of pleiotropic effects. The real challenge, however, is to determine how evolutionary and developmental modules relate to each other. Intuitively, developmental and evolutionary modules should be closely related. The developmental process determines how a gene influences the phenotype, and hence the existence of developmental modules should influence the structure of the genotype-phenotype map. Developmental modules, however, can be deployed repeatedly, as in the case of the left and right forelimb bud. According to a current explanation, each of the two forelimb buds is an independent developmental module, because each is a self-contained developmental unit with its own capacity for self-differentiation. From a variational point of view, however, the left and right forelimbs are not independent, because they express the same genetic information. Mutations are thus expected to affect both forelimbs simultaneously, and the genetic variations of the two limbs are correlated. Hence, the two forelimbs are two *different developmental modules* of the organism and are also parts of the *same evolutionary module*. The existence of developmental modules, on the other hand, may play a role in the origin of evolutionary modules. Proceeding further, according to Gilbert and Bolker (2001), there are signal transduction pathways that integrate embryonic development and constitute modules that can be considered homologous, in the same way in which structures can be considered homologous. Such homologies of processes are a critical issue, because evolution depends on heritable changes in development, where different modules can change without affecting other modules and can be co-opted into new functions.

A central point is how these units emerge. Wagner (1996) argued that homologues can be understood as modular units of evolutionary transformation and that they may appear spontaneously by self-organization or may be the product of natural selection. A selection scenario that could explain the origin of modular units needs to explain the differential suppression of pleiotropic effects between different modules, and the augmentation of pleiotropic effects among the elements within a single module. After considering different scenarios, Wagner concluded that a combination of directional and stabilizing selection is a prevalent mode of selection and a likely explanation.

If this connection between development, modules, and homologues is accepted, the question now arises of how and to what extent it affects the modules envisaged in evolutionary psychology.

As thoroughly discussed by Cosmides and Tooby (Cosmides and Tooby 1992; Tooby and Cosmides 1995), according to evolutionary psychology, human neural circuits were designed by natural selection to solve problems that our ancestors faced during the evolutionary history of our species. In addition, different neural circuits are specialized for solving different adaptive problems. On these tenets, the adaptationist approach to psychology searches for an adaptive design, which usually entails the consideration of niche-differentiated mental abilities, unique to the species under examination. This *adaptationist approach* is opposed to the classical *phylogenetic approach*, which rests on the search for phylogenetic continuities, due to the inheritance of homologous features from common ancestors. This latter way of reasoning programmatically overlooks homology in evolutionary psychology. Even the former, adaptationist approach, though, while strongly

61

evoking modularity, limits its interest for modules to the search for problem-solving circuits. Furthermore, evolutionary psychology must also face many criticisms (Bjorklund and Pellegrini 2003) on purely psychological grounds. Modularity is questioned by developmental psychologists, who believe that modules are not static but "grow" during a child's development (Karmiloff-Smith 1992). In addition, the idea of mapping the mind (Hirschfeld and Gelman 1994) poses severe difficulties, because it requires the tricky definition of precise boundaries for the domain of knowledge and for cognitive modules.

Homology and the Study of Brain and Behavior

In classical ethology, homology had a core position. Eibl-Eibesfeldt (1970), for instance, wrote that homologous "behavior patterns are of great taxonomic value and can help to elucidate the natural relationships among animals." Moreover, he observed that "descent in most cases implies a direct genetic relationship, where the information, which concerns the adaptiveness of the behavior pattern in question, is passed through the genome."

Behavioral ecology, on the contrary, was not particularly interested in observing behavioral homologies but rather in investigating convergent adaptations (i.e., analogies). The principal aim of this comparative approach was to compare the associations between ecology, evolution, and behavioral traits, with an emphasis on their possible adaptive consequences at an individual level. In contrast with classical ethology, however, only gross behavioral and ecological traits were compared in order to formulate laws describing causal or regularity relations among variables (Jarman 1981; Krebs and Davies 1997).

Finally, in many cases, the similarities between characters at one level (e.g., behavioral) were based on evidences suitable for a different level. An example might be Hodos's definition (1976): "Behaviors are considered homologous to the extent that they can be related to specific structures that can, in principle, be traced back through a genealogical series to a stipulated ancestral precursor irrespective of morphological similarity." Thus the primacy returns to brain structure, and its homologues.

During the past 150 years, comparative neurology accumulated an enormous load of data and hypotheses on brain changes in different taxa (for vertebrates, cf. Butler and Hodos 1996; Nieuwenhuys 1998; Roth and Wullimann 2001). These studies, well set in a naturalistic and comparative outlook, had very different aims. In many cases, variations of brain structures were used to infer taxonomic relations by means of a cladistic approach (cf. Northcutt 1984). In other studies, brain changes driven by highly specific niche adaptation or by special sensory modalities were investigated (Kaas 2000;

Catania 2000). Finally, patterns of brain variation were used to elaborate theories of the mechanisms for change, at the morphological and/or molecular level.

In all these studies, homology and the related concepts were fundamental, even if the definition of characters to compare was highly debated and frequently reviewed, along with the development of new technical approaches (Fasolo and Malacarne 1988). The criteria to infer homology moved from gross anatomy, to pathways comparison, to neurochemical mapping, to genetic expression patterns (Nieuwenhuys 1998; Roth and Wullimann 2001), involving mainly adult mapping characters. Recently, the interest has focused on extrapolations from structures to functions, in virtue of the fact that there are now available extended databases on the central nervous system, as well as databases on molecular biology. *NeuroHomology*, for instance, considers interrelated sets of data on brain structures, neural connections, and homologies in different species. It relies on a set of rules that are embedded in the database and are supposed to deliver a relatively objective evaluation of both neuroanatomical connections and the degree of homology (Bota and Arbib 2001).

Brain homology and function remain, however, in an "uneasy alliance" (Striedter 2002).

Recent results in molecular genetics highlight a common genetic regulatory background for the central nervous system (CNS), shared by both vertebrates and invertebrates (Hirth and Reichert 1999; Fritzsch 1998): regulatory genes and extrinsic mechanisms control the realization of the neocortex (O'Leary and Nakagawa 2002). In general, however, comparative studies on brain evolution face the clash between the foreseeable stability of the basic proposal and the huge degree of variation, even at species-specific level. The recent research, which results from the blooming of neurobiological approaches, focuses on the neuronal circuits underlying species-specific behavior and the genes it involves (Katz and Harris-Warrick 1999). On the other hand, many recent studies on the CNS, particularly on the neocortex, are stressing the occurrence of homologues within different mammalian taxa (Krubitzer 1995) and even in common with the avian pallium, called Wulst (Medina and Reiner 1999), as well as evolutionary novelties (Northcutt and Kaas 1995).

Theories of CNS Evolution

While reviewing our changing views on brain evolution, Northcutt (2001) suggested that the rapprochement of embryology and genetics is fueling a new renaissance, which promises to increase our understanding of brain evolution and of its genetic basis. Thus, he seems to be changing his previously skeptical views, according to which the "Critical evaluation of current hypotheses concerning CNS evolution reveals that these hypotheses generally describe patterns of character variation and rarely address processes" (Northcutt 1984, p. 712).

One of the reasons for this radical change of view is the huge amount of new data arising from molecular developmental genetics, but mainly from some new approaches, which manage to cope with the developmental and epigenetic processes leading to species-specific, or even individual, brains, through an account of the organization of plastic neural matter. These theories include somatic selectionisms (e.g., neural Darwinism: Edelmann 1987, 1988, 1993), the theory of the elective stabilization of synapses (Changeux 1983, 2002), neurotrophic theory (Purves 1994), axonal displacement (Deacon 1990, 1995, 1997), and rather different (sometimes even contrasting) views about the mechanisms of evolution. They all, though, stress the central role of selective mechanisms, which act positively or negatively during the development of an individual and thus manage to accustom its brain to internal and external environments. From these points of view, epigenetic mechanisms are among the steps that are necessary to accomplish adult brain regularities through ontogeny. As a result, the genetic blueprint can be acted out by intercellular, environmental, and experiential communication. The upshot is that selection mechanisms are not simply direct shaping processes responding to old or new pressures, or the exploitation of a niche. Plasticity in brain connections, after changes of the visual afferences to the cortex, represents a beautiful example of Evo-Devo mechanisms and gives room to the concept of epigenetic homology (Striedter 1998).

Another intriguing point is represented by the so-called embryonic fields (Redies and Puelles 2001; Puelles and Medina 2002). Embryonic fields are histogenetic modules that are specified by the position-dependent expression of patterning genes. Within each embryonic module, secondary and higher-level pattern formations take place during development and, at the end, give rise to brain nuclei and cortical layers. Defined subsets of these structures become connected by fiber tracts and form the information-processing neural circuits, which represent the functional modules of the brain. Embryonic modularity is transformed into functional modularity, in part by translating early-generated positional information into an array of adhesive cues, which regulate the binding of functional neural structures distributed across the embryonic modules. Such brain modularity may provide a basis for evolvability.

The Quantitative Approach

The other leading approach to brain evolution is the quantitative analyses of relative brain size, or structural encephalization, which has the purpose of finding biological correlates of minds in animals. Following the seminal work by Jerison (1973) on the quantitative changes of brain size in vertebrates, some studies were carried out to evaluate the brain growth in some groups, notably cetaceans and primates. These studies originated the concept of encephalization, in reference to animals, the brains of which are over the expected brain-body allometric curve. The question is: how dependent is brain size on body size? How dependent is it from other factors? Jerison's concept of "extra" cortical neurons follows a long tradition of attempts to divide brain size into a fraction that is supposed to be necessary for somatic maintenance (i.e., dependent on body size) and a fraction that reflects actual encephalization (i.e., the neurons developed in order to deal with extracorporeal pressures, e.g., somatic and psychic brain functions). In his view, the "extra" neurons are a by-product of the encephalization process, which makes new behavioral mechanisms possible; these are relatively unusual kinds of behavior, which require a neural information-processing capacity larger than that which can be explained by differences in body size across different species. From this perspective, the different kinds of behavior, which depend on an augmented processing capacity, are evidence of the emergence of different kinds of intelligence throughout evolution (cf. Jerison 1991).

Recent quantitative evaluation led to some contrasting views. Overall, there seem to be two broad models for how brains change. On the one hand, their parts might be taken to be distinguishable in virtue of their functions and to vary independently, one from the other. Brain evolution, in that case, would be a matter of growing a bigger auditory processing system, for example, while the other systems remain mostly unchanged. Thus Barton and Harvey (2000) argued, on the basis of quantitative data, that the natural selection of particular behavioral capacities might selectively cause size changes of the systems running those capacities. Their comparative data support the idea that this sort of "mosaic" evolution has been an important factor in the evolution of brain structure, because the neocortex shows about a fivefold difference in volume between primates and insectivores, even after accounting for its scaling relationship with the rest of the brain. In addition, brain structures with major anatomical and functional links evolved together independently of evolutionary changes in other structures. This is probably true at the level of both basic brain subdivisions and more fine-grained functional systems.

Hence, brain evolution in these groups involved complex relationships among individual brain components.

Alternatively, the size of the entire brain might be taken to vary in response to the selection of any of its constituent parts. In this model, constraints depending on the developmental structure of the brain condition the dimensional proportions of its parts and, as a result, limit its aptitude to respond to selection. Finlay et al. (2001), for instance, concluded that any substantial change in brain size requires a change in the number of neurons and their supporting elements, which in turn requires an alteration in either the rate or the duration of neurogenesis. The schedule of neurogenesis is surprisingly stable in mammalian brains, and increases in the duration of neurogenesis have predictable outcomes: late-generated structures become disproportionately large. The olfactory bulb and associated limbic structures may deviate in some species from this general pattern of growth: in rhesus monkeys, the size reduction of the limbic system appears to be produced by an advance in the onset of terminal neurogenesis in limbic system structures. Besides neurogenesis, many other cases of neural maturation, such as process extension and retraction, follow the same schedule and manifest the same degree of predictability. Although the underlying order of events remains the same for all of the mammals so far studied, related subclasses (e.g., marsupial and placental mammals) differ in the overall rate of neural maturation; even within the same subclass, different species differ in the duration of neurodevelopment. A substantial part of the regularities of event sequences in neurogenesis can be related directly to the two dimensions of the neuraxis. Both the spatial and the temporal organization of development have been highly conserved throughout mammalian brain evolution and represented a strong constraint on the types of possible brain adaptations. The neural mechanisms of integrative behavior may be found in those locations which, for the number of neurons, have enough plasticity to support them. According to theories of this sort, changes of size might not be selected for but might represent a by-product of some developmental constraints and – *a posteriori* – permit new behavioral repertoires and adaptation clues. The theoretical clue is the role of adaptation in driving brain changes. Many relevant features such as an enlarged isocortex, according to Finlay et al. (2001), might be a "spandrel, a by-product of structural constraints later adapted for various behaviors, in contrast with the diffusely maintained idea that some particular regions are selected for cognitively advanced uses."

Through a comparative account of human neuroanatomy and development, Deacon (2000) connected his views on somatic selectionism to quantitative changes. He recognized a number of ways in which human brains diverge from the general pattern of other ape and monkey brains. These divergences may offer hints about language evolution. Analyses of large-scale quantitative changes in the relative proportions of brain regions (as opposed to the analyses of mere overall expansion) offer some of the most obvious clues. Additional information about how axons are guided in their extensions to distant developmental targets and how competitive trophic processes sculpt these connections also provides a way to understand how gross quantitative changes in cell numbers could affect circuit organization and, ultimately, behavior.

Interestingly enough, while reviewing theories of the evolution of brain and intelligence in primates, Rifkin (1995) related grades in encephalization to different needs (metabolic influence, foraging, social complexity and Machiavellian intelligence, group size and social structure). His conclusions were: "There *are* undeniable trends in the history of life – towards larger brains in mammals and larger neocortices in primates – but to generalize correlations of these trends into a concept of intelligence should not be attempted until an accurate definition is developed. Until that time, the most that comparative brain size studies can do is to demonstrate correlations and thereby pose questions for scientists who focus on the evolution of species with one of these correlated characteristics."

Evolutionary Reasoning: Adaptation, Exaptation, Coevolution

The comparative method can provide the evolutionary analysis of brain and cognition with powerful insights (Roth and Wullimann 2001). From the standpoint of this comparative approach, the opposition between phylogenetic and adaptationistic approaches – as assumed in current scientific practice – seems historically untenable. Studies in cognitive neuroscience, for example, those which stress the diversity among mammalian brain organization (Preuss 1995) and cognitive specializations (cf. Povinelli and Preuss 1995), are fully compatible with the idea that, in primates, the neocortex shares homologies with that of other mammals.

Even the "uneasy alliance," which is implicit in expressions like "functional homology" as used in some recent papers (cf. Rizzolatti et al. 2002 about monkey area F5 and human area 44), can be strengthened by a comparative program for the search of homologies. Brain mapping in primates and humans, through new imaging technologies, gives powerful means to compare particular areas (e.g., visual areas) and reveals many intraspecific similarities as well as striking differences (Sereno 1998).

The trendy interest in development can effectively enrich our definitions of homology and our methods to individuate it. The study of developmental processes calls for a comparison at different developmental stages, overcoming the restriction to adults, which has been the focus in classical comparative studies. Too often, comparative neurobiologists have considered brain evolution as the transformations of adult brains over time (Northcutt 2001). A more extensive interest in dynamic processes can help unveil the plastic changes of the brain throughout life. To give a simple example, the developing human brain seems to be different at the functional neuroanatomical level from the adult brain, even in processing single words (Schlaggar et al. 2002). In order to cope with these complex behavioral patterns of a developing organism, new theoretical approaches are currently underway. The combinatory approach (Minelli 1998) and the hierarchical approach (Abouheif 1997) are two examples. Recently, the concept of partial homology also emerged (Minelli 2003). According to Wake, the concept of partial homology depends on the very idea of evolutionary change: "Because evolution is a continuous process, . . . homology can be only ever partial, in any real sense" (1999, pp. 44–45).

Another puzzling problem is the genesis of novelty and its adaptive value. Interestingly enough, very recent molecular investigation on primates shows that the human brain has probably experienced pronounced evolutionary changes in gene expression during its most recent history (Enard et al. 2002). These results are open to different theoretical interpretations, but they suggest that processes of fast genetic reorganization might sometimes occur.

In light of these considerations, one main question emerges: what are the adaptive pressures behind brain and behavior novelties in evolution? We have no answer yet, but we can agree with the original statement by Williams, in his 1966 *Adaptation and Natural Selection*, frequently quoted in evolutionary psychology, but not so frequently exploited: "*Evolutionary adaptation is a special and onerous concept that should not be used unnecessarily, and an effect should not be called a function unless it is clearly produced by design and not by chance. When recognized, adaptation should be attributed to no higher a level of organization than is demanded by the evidence*" (emphasis added).

A new emphasis on homology in evolutionary biology (the persistence of theoretical problems notwithstanding) may offer new powerful tools for an effective comparative analysis, and may thus help distinguish between strict biological correspondence and loose metaphoric representations of behavior, which are the mere result of an uncritical assumption of an evolutionary stance. Especially in cases of highly complex behavior, ethics being a paradigmatic example, biology and culture are certainly tightly entrenched: the claim that these kinds of behavior have evolutionary bases is simply a truism. The evolution of the brain involved a complex set of relationships among individual structures, both at the quantitative and the qualitative level. As aforementioned, there is some controversy concerning this idea, but the core problem (e.g., whether changes are directly selected or not) remains unsolved. It seems plausible, however, that some processes are related to environmental pressures, while others emerge in response to the need for more flexible answers, and still others are part of a less specific and foreseeable ecological niche. Likewise, brain structures have developed along several lines, and one usually finds a "mosaic-like" pattern even within a particular line. In other words, an animal may have a high degree of specialization or efficiency in some brain areas or behavioral patterns, although this might not be the case for other parts (cf. Fasolo and Malacarne 1988). This view forces us to avoid quick generalizations. Any specific circuit or behavior has to be investigated with comparative tools, in order to assess the entrenchment of continuity and novelty.

In such a mosaic of integrated parts, whatever the evolutional process might have been, a large part of the variation has not been selected per se, but represents a collection of exaptations (Gould 2002).

The idea that novelty may arise from an exaptation process has a high impact on our views of evolutionary trends and supports, at least in principle, Boniolo's conclusion, that is, that moral capacity has to be considered as an evolutionary outcome that resulted from changes of suitable cerebral traits allowing its enabling conditions (Boniolo, Chapter 2 in this volume).

REFERENCES

Abouheif, E. 1997. Developmental Genetics and Homology: A Hierarchical Approach. *Trends Ecol. Evol.* 12: 405–408.

Barton, R. A., and Harvey, P. H. 2000. Mosaic Evolution of Brain Structure in Mammals. *Nature* 405: 1055–1058.

Bateson, W. 1882. On Numerical Variation in Teeth with a Discussion of the Conception of Homology. *Proc. Zool. Soc.* London: 102–115.

Bateson, W. 1884. Materials for Study of Variation. Reprint, Baltimore: Johns Hopkins University Press, 1992.

Bjorklund, D. F., and Pellegrini, D. A. 2003. *The Origin of the Human Nature: Evolutionary Developmental Psychology.* Washington, D.C.: American Psychological Association.

Bota, M., and Arbib, M. A. 2001. The Neurohomology Database. *Neurocomputing* 38–40: 1627–1631.

Bullock, T. H. 1983. Implications for Neuroethology from Comparative Neurophysiology. In *Advances in Vertebrate Neuroethology*, ed. J. P. Ewert, R. R. Caprinica, and D. J. Ingle, 53–75. New York: Plenum.

Butler, A. B., and Hodos, W. 1996. *Comparative Vertebrate Neuroanatomy: Evolution and Adaptation.* New York: Wiley-Liss.

Butler, A. B., and Saidel, W. M. 2000. Defining Sameness: Historical, Biological and Generative Homology. *BioEssays* 22: 846–853.

Catania, K. C. 2000. Cortical Organization in Insectivora: The Parallel Evolution of Sensory Cortex and the Brain. *Brain Behavior and Evolution* 55: 311–321.

Changeux, J. P. 1983. *L'homme neuronal.* Paris: Fayard.

Changeux, J. P. 2002. *L'homme de verité.* Paris: Odile Jacob.

Cosmides, L., and Tooby, J. 1992. Cognitive Adaptations for Social Exchange. In *The Adapted Mind,* ed. J. Barkow, L. Cosmides, and J. Tooby, 163–228. Oxford: Oxford University Press.

Darwin, C. 1871. *The Descent of Man.* Amherst, N.Y.: Prometheus Books, 1998.

de Beer, G. R. 1971. *Homology, an Unsolved Problem.* Oxford: Oxford University Press.

Deacon, T. W. 1990. Rethinking Mammalian Brain Evolution. *Amer. Zool.* 30: 629–705.

Deacon, T. W. 1995. On Telling Growth from Parcellation in Brain Evolution. In *Behavioural Brain Research in Naturalistic and Seminaturalistic Settings: Possibilities and Perspectives*, ed. E. Alleva, H.-P. Lipp, A. Fasolo, L. Nadel, and L. Ricceri, 37–72. Dordrecht: Kluwer.

Deacon, T. W. 1997. *The Symbolic Species: The Co-evolution of Language and the Brain.* New York: W. W. Norton.

Deacon, T. W. 2000. Evolutionary Perspectives on Language and Brain Plasticity. *J. Commun. Disord.* 33: 273–291.

Edelman, G. M. 1987. *Neural Darwinism: The Theory of Neuronal Group Selection.* New York: Basic Books.

Edelman, G. M. 1988. *Topobiology: An Introduction to Molecular Embryology.* New York: Basic Books.

Edelman, G. M. 1993. Neural Darwinism: Selection and Reentrant Signaling in Higher Brain Function. *Neuron* 10: 115–125.

Eibl-Eibesfeldt, I. 1970. *Ethology: The Biology of Behavior.* New York: Holt & Rinehart.

Enard, W., et al. 2002. Intra- and Interspecific Variation in Primate Gene Expression Patterns. *Science* 296: 340–343.

Fasolo, A., and Malacarne, G. 1988. Comparing the Structure of Brains: Implications for Behavioral Homologies. In *Intelligence and Evolutionary Biology*, ed. H. Jerison and I. Jerison, 119–142. Berlin: Springer-Verlag.

Finlay, B. L., Darlington, R. B., and Nicastro, N. 2001. Developmental Structure in Brain Evolution. *Behav. Brain. Sci.* 24: 263–278.

Fritzsch, B. 1998. Of Mice and Genes: Evolution of Vertebrate Brain Development. *Brain. Behav. Evol.* 52(4–5): 207–217.

Gilbert, S. F., and Bolker, J. A. 2001. Homologies of Process and Modular Elements of Embryonic Construction. *J. Exp. Zool.* 291(1): 1–12.

Goodwin, B. C. 1984. Changing from an Evolutionary to a Generative Paradigm. In *Evolutionary Theory: Paths into the Future,* ed. J. W. Pollard, 99–120. New York: Wiley.

Gould, S. J. 1977. *Ontogeny and Phylogeny.* Cambridge, Mass.: Harvard University Press.

Gould, S. J. 2002. *The Structure of the Evolutionary Theory.* Cambridge, Mass.: Harvard University Press.

Hirschfeld, L., and Gelman, S. 1994. *Mapping the Mind: Domain Specificity in Cognition and Culture.* Cambridge: Cambridge University Press.

Hirth, F., and Reichert, H. 1999. Conserved Genetic Programs in Insect and Mammalian Brain Development. *BioEssays* 21: 677–684.

Hodos, W. 1976. The Concept of Homology and the Evolution of Behavior. In *Evolution, Brain and Behavior: Persistent Problems*, ed. R. B. Masterton, W. Hodos, and H. J. Jerison, 153–167. Hillsdale, N.J.: Lawrence A Erlbaum Associates.

Houle, D. 2001. Characters as the Units of Evolutionary Change. In *The Character Concept in Evolutionary Biology*, ed. G. P. Wagner, 109–140. San Diego: Academic Press.

Jarman, P. J. 1981. Prospects for Interspecific Comparison in Sociobiology. In *Current Problems in Sociobiology,* ed. King's College Sociobiology Group, Cambridge, 323–342. Cambridge: Cambridge University Press.

Jerison, H. J. 1973. *Evolution of the Brain and Intelligence.* New York: Academic Press.

Jerison, H. J. 1991. *Brain Size and the Evolution of Mind.* The 59th James Arthur Lecture on the Evolution of the Human Brain. American Museum of Natural History, New York.

Kaas, J. H. 2000. Organizing Principles of Sensory Representations. In *Evolutionary Developmental Biology of the Cerebral Cortex*, ed. G. R. Bock and G. Cardew. *Novartis Foundation Symposium* 228: 188–189. Chichester: Wiley.

Karmiloff-Smith, A. 1992. *Beyond Modularity: A Developmental Perspective on Cognitive Science.* Cambridge, Mass.: MIT Press.

Katz, P. S., and Harris-Warrick, R. M. 1999. The Evolution of Neuronal Circuits Underlying Species-Specific Behavior. *Curr. Opin. Neurobiol.* 9(5): 628–633.

Kim, J., and Kim, M. 2001. The Mathematical Structure of Characters and Modularity. In *The Character Concept in Evolutionary Biology,* ed. G. P. Wagner, 215–236. San Diego: Academic Press.

Krebs, J. R., and Davies, N. B. 1997. *Behavioural Ecology: An Evolutionary Approach.* Oxford: Blackwell.

Krubitzer, L. 1995. The Organization of Neocortex in Mammals: Are Species Differences Really So Different? *Trends in Neuroscience* 18: 408–417.

Medina, L., and Reiner, A. 1999. Do Birds Possess Homologues of Mammalian Primary Visual Somatosensory and Motor Cortices? *Trends in Neuroscience* 23: 1–12.

Miller, P. H. 1996. "Mapping the Mind": Where Are the State Lines? *Cogn. Devel.* 11: 141–155.

Minelli, A. 1998. Molecules, Developmental Modules, and Phenotypes: A Combinatorial Approach to Homology. *Molec. Phylog. Evol.* 3: 340–347.

Minelli, A. 2003. *The Development of Animal Form.* Cambridge: Cambridge University Press.

Müller, G. B., and Newman, S. A. 1999. Generation, Integration, Autonomy: Three Steps in the Evolution of Homology. In *Homology,* ed. G. R. Bock and G. Cardew. *Novartis Foundation Symposium* 222: 65–79. Chichester: Wiley.

Nieuwenhuys, R. 1998. Comparative Neuroanatomy: Place, Principles and Program. In *The Central Nervous System of Vertebrates*, ed. R. Nieuwenhuys, H. J. Ten Donkelaar, and C. Nicholson, 273–326. Berlin: Springer-Verlag.

Northcutt, R. G. 1981. Evolution of the Telencephalon in Nonmammals. *Ann. Rev. Neurosci.* 4: 301–350.

Northcutt, R. G. 1984. Evolution of the Vertebrate Central Nervous System: Patterns and Processes. *Amer. Zool.* 24: 701–716.

Northcutt, R. G. 2001. Evolution of the Nervous System. Changing Views of Brain Evolution. *Brain Research Bulletin* 55: 663–674.

Northcutt, R. G., and Kaas, J. H. 1995. The Emergence and Evolution of Mammalian Neocortex. *Trends in Neuroscience* 18: 373–379.

O'Leary, D., and Nakagawa, Y. 2002. Patterning Centers, Regulatory Genes and Extrinsic Mechanisms Controlling Arealization of the Neocortex. *Curr. Opin. Neurobiol.* 12: 14–25.

Patterson, C. 1982. Morphological Characters and Homology. In *Problems of Phylogenetic Reconstruction,* ed. K. A. Joysey and A. E. Frida, 21–74. New York: Academic Press.

Povinelli, D. J., and Preuss, T. M. 1995. Theory of Mind: Evolutionary History of a Cognitive Specialization. *Trends in Neuroscience* 18: 418–424.

Preuss, T. M. 1995. The Argument from Animals to Humans in Cognitive Neuroscience. In *The Cognitive Neurosciences,* ed. M. Gazzaniga, 1127–1241. Cambridge, Mass.: MIT Press.

Puelles, L., and Medina, L. 2002. Field Homology as a Way to Reconcile Genetic and Developmental Variability with Adult Homology. *Brain Research Bulletin* 57: 243–255.

Purves, D. 1994. *Neural Activity and the Growth of the Brain.* Cambridge: Cambridge University Press.

Redies, C., and Puelles, L. 2001. Modularity in Vertebrate Brain Development and Evolution. *BioEssays* 23: 1100–1111.

Rifkin, S. 1995. The Evolution of Primate Intelligence. *Harvard Brain,* vol. 2. http://hcs.harvard.edu/~husn/BRAIN/vol2/Primate.html.

Rizzolatti, G., Fogassi, L., and Gallese, V. 2002. Motor and Cognitive Functions of the Ventral Premotor Cortex. *Curr. Opin. Neurobiol.* 12: 149–154.

Roth, G., and Wullimann, M. (eds.). 2001. *Brain Evolution and Cognition.* New York: Wiley.

Schlaggar, B. L., et al. 2002. Functional Neuroanatomical Differences between Adults and School-Age Children in the Processing of Single Words. *Science* 296: 1476–1479.

Sereno, M. I. 1998. Brain Mapping in Animals and Humans. *Curr. Opin. Neurobiol.* 8: 188–194.

Striedter, G. F. 1998. Stepping into the Same River Twice: Homologues as Recurrent Attractors in Epigenetic Landscapes. *Brain Behav. Evolution* 38: 177–189.

Striedter, G. F. 2002. Brain Homology and Function: An Uneasy Alliance. *Brain Research Bulletin* 57: 239–242.

Tooby, J., and Cosmides, L. 1995. Mapping the Evolved Functional Organization of Brain and Mind. In *The Cognitive Neurosciences,* ed. M. Gazzaniga, 1185–1198. Cambridge, Mass.: MIT Press.

van Valen, L. M. 1982. Homology and Causes. *J. Morphol.* 173: 305–312.

Waddington, C. 1957. *The Strategy of the Genes.* London: Allen and Unwin.

Wagner, G. P. 1994. Homology and the Mechanisms of Development. In *Homology: The Hierarchical Basis of Comparative Biology*, ed. B. K. Hall, 274–301. New York: Academic Press.

Wagner, G. P. 1996. Homologues, Natural Kinds and the Evolution of Modularity. *Am. Zool.* 36: 36–43.

Wagner, G. P., and Altenberg, L. 1996. Complex Adaptations and the Evolution of Evolvability. *Evolution* 50: 967–976.

Wagner, G. P., Mezey, J., and Calabretta, R. 2005. Natural Selection and the Origin of Modules. In *Modularity: Understanding the Development and Evolution of Complex Natural Systems*, ed. W. Callabaut and D. Rasskin-Gutman, 33–49. Cambridge, Mass.: MIT Press.

Wake, D. B. 1994. Comparative Terminology. *Science* 265: 268–269.

Wake, D. B. 1999. Homoplasy, Homology and the Problem of "Sameness" in Biology. In *Homology,* ed. G. R. Bock and G. Cardew. *Novartis Foundation Symposium* 222: 24–46. Chichester: Wiley.

West-Ebehard, M. J. 2003. *Developmental Plasticity and Evolution*. Oxford: Oxford University Press.

Wiley, E. O. 1981. *Phylogenetics: The Theory and Practice of Phylogenetic Systematics*. New York: Wiley.

Williams, G. C. 1966. *Adaptation and Natural Selection: A Critique of Some Current Evolutionary Thought*. Princeton: Princeton University Press.

Wray, G. A. 1999. Evolutionary Dissociations between Homologous Genes and Homologous Structures. In *Homology,* ed. G. R. Bock and G. Cardew. *Novartis Foundation Symposium* 222: 189–203. Chichester: Wiley.

III

How Biological Results Can Help Explain Morally Relevant Human Capacities

III

5

Genetic Influences on Moral Capacity

What Genetic Mutants Can Teach Us

GIOVANNI BONIOLO AND PAOLO VEZZONI

> Lo maggior don che Dio per la sua larghezza
> fesse creando, ed a la sua bontate
> più conformato, e quel ch'è più apprezza,
> fu de la volontà la libertate;
> di che le creature intelligenti,
> e tutte e sole, furono e sono dotate.
> Dante Alighieri, *Paradiso*, V, 19–24

Discussions about the relations between genes and free will, in particular in relation to morality, are nothing but a new chapter of an old story, the beginning of which can be traced back before the birth of Western philosophy. Is man really free in choosing moral values and moral actions? Are there biological, physical, metaphysical, or divine constraints?

We know that throughout the history of philosophy, the debates on moral freedom have reached an extremely high degree of sophistication. Certainly this is not the place to deal with this vast topic. More modestly we will limit ourselves to consider some data coming from genetics, in order to understand, from a biological point of view, to what extent we can say we have free moral will, and how far our freedom can reach.

A long-standing tradition, well expressed in Dante Alighieri's verses, holds the view that free will is a specific characteristic of humans. But some scholars thought that all human choices are determined. In particular there were those supporting the idea that man's choices are determined by his genome: we would be our genes. We know, for example, what the Nobel laureate James Watson said: "Now we know, in large measure, our fate is in our genes" (cf. Jaroff 1989). And Gilbert (1992) observed that, because our sequenced

We dedicate this essay to Paolo Raineri, physician and maestro of philosophy.

77

genome can be contained in a CD, we may take the latter out of our pocket and claim: "Here is a human being: this is me!" If we were our genome, our choices would be written in it, and all the claims about our free will would be mere *flatus vocis*: what might appear as free choices actually would be reduced to genetic mutations, recombinations, and rearrangements. Dante notwithstanding.

Indeed, there were several molecular biologists who supported genetic determinism, but there were, and there are, also several biologists believing that our behavior, in particular our ethically assessable behavior, is *Not in Our Genes* (Lewontin et al. 1984; cf. also Lewontin 1991; Berkowitz 1996).[1] To be honest, at the present time, we suspect that no biologically informed and philosophically critical person could rationally support a strong form of genetic determinism.[2] Now we know that not everything is in our genes. Nevertheless, we cannot overlook the fact that *something is in our genes*. The problem, then, turns into understanding what is in our genes, and to what extent that can constrain, or even determine, our free moral valuing and acting.

This is a truly important issue, from both a philosophical and a biological point of view, and in what follows we attempt to highlight this issue, that is, what "something is in our genes" means. We do this by taking into great account Locke's suggestion: "it becomes the Modesty of Philosophy, not to pronounce Magisterially, where we want that Evidence that can produce Knowledge; . . . that it is of use to us, to discern how far our Knowledge does reach" (Locke 1690, book IV, chap. III, § 7, pp. 541–542). Therefore we pronounce philosophically only when biological evidence can support our claim. In particular, accepting Boniolo's (Chapter 2 in this volume) suggestions that there is a difference between behavior, moral judgments on behavior, and moral capacity, we focus our attention on the possible genetic influences on moral capacity, intended "as the capacity both for formulating and applying moral judgments on behavior, and for behaving accordingly." We accept also the idea, there argued, according to which only the enabling conditions for the moral capacity can be investigated biologically; but moral systems cannot. This means that if there are biological enabling conditions for the moral capacity and if they are defective, or "nonnormal," then also the moral capacity should be defective, or "nonnormal."

---

[1] With reference to Lewontin, it should be noted that his motto is also a result of an ideological criticism. On this question, cf. Vezzoni 2000.

[2] However, if there were still some "Japanese soldier," unaware that the war is over and that it has been lost, we might always investigate the reasons why he holds such a position. Cf. Oyama 1985; Kitcher 2001.

In what follows we work exactly along this line, and we investigate in which sense, in some deviant cases, an agent's moral capacity is genetically influenced. This discussion may help explain what "something in our genes" means. Of course, how these deviant cases should be morally assessed does not depend on genetics, but on one's moral theory, in particular on the account of moral responsibility: genetics does not at all offer the grounds for a moral assessment of behavior.

GENETIC INFLUENCES

"Is the behavioral phenotype influenced by the genotype?" Unfortunately, like any other living beings on Earth, we humans cannot give a totally exhaustive answer. However, something can be said about these influences, even if we must be extremely careful, since we do not yet know enough (cf. also de Belle 2002).

It is a platitude that the behavior of rabbits and wolves is influenced by their genes; it is quite rare to observe a rabbit behaving like a wolf, or a wolf behaving like a rabbit. A mouse and a bird, if subjected to the same conditioning learning program, will acquire two extremely different sorts of behavior: their learning capacities are extremely different, and this is also due to their genetic differences (cf. Garcia et al. 1972). Therefore there should be no doubts that behavior is genetically influenced. However, it is also true that the role of learning should not be underestimated either: even wolves undergo training, and a strong learning program on young monkeys has shown a significant influence on their innate sexual behavior (cf. Harlow et al. 1965). Moreover, a role is also played by contingent conditions: a wolf with a full stomach does not behave like a wolf with an empty stomach.

We know that each gene, that is, a given encoding nucleotide sequence, occupies a certain place in a given chromosome. We know also that (almost) any given gene codes for an amino acid sequence, that is, for a protein, can be considered the phenotypic result. More generally, however, when we speak about phenotypic traits, we are not speaking about proteins but about more visible traits, such as eye color, size, and some sorts of behavior. There are monogenic phenotypic traits, depending on the expression of one gene only, and polygenic phenotypic traits, depending on the expression of more than one gene. In what follows we consider some *monogenic and polygenic genetic diseases*, that is, diseases that occur because the "right" nucleotide sequence of one or more genes – respectively – has changed. Through this review, we will see how there are genetic influences on the moral capacity. But, *be alert!* We should be careful not to generalize these results more than they can be ("it becomes the Modesty of Philosophy ...").

We analyze pathological cases arising both without any biologist's intentional intervention and with its intentional intervention concerning the turning on, the turning off, or the changing of the structure of certain genes. Needless to say, this last kind of intervention has not been carried out on humans but only on animals, mice in particular.

Another aspect worth remarking regards polygenic traits. Some of them can depend both on the interactions of several genes and on other extragenic factors. A normal blood glucose level depends on many factors, such as the hormones causing it to increase and decrease; but it also depends on food, on the level of physical exercise, and other factors. Pathology, however, remains the best-known condition in which to study polygenic traits. Over the past few years, we have realized that polygenic diseases very rarely depend on the complete inactivation of singular genes, due to some severe changes; usually, they depend on small variations, slightly modifying the functions of genes. Moreover, two conclusions can be drawn from the genetic analysis of the most frequent complex diseases, such as diabetes, hypertension, obesity, and osteoporosis. First, each gene contributes only a small part to the pathogenesis of the disease. Second, variations in these genes are not an all-or-nothing matter, as happens in the case of monogenic diseases, but rather they are a matter of degree. A gene may not be completely inactivated but simply slightly under- or overexpressed; or its structure may be altered in a way such that the protein produced by it ends up working in a slightly unusual way. The molecular basis of this phenomenon is genetic polymorphism, namely the fact that the nucleotide sequence in specific points of the genome is not unique.

The task of identifying the polymorphisms, which are responsible for complex diseases, is not a trivial one, even with current techniques. Indeed, so far very few positive association studies have survived careful check and been successfully reproduced by independent researchers (cf. Risch 2000). Moreover, the study of morphological phenotypic traits (e.g., height) is even trickier, and very few data are available in this field. Skin color is a clear example: it is certainly a genetically inherited trait, but its molecular basis is still unknown.

In short, at one extreme there are monogenic traits, and therefore cases in which a malfunctioning of a specific gene can be quite straightforwardly associated with a specific disease. At the other extreme, there are strongly polygenic traits (such as the psychological traits), and in these cases the correct and precise gene dependence (and, thus, the hereditability) is still hypothetical. In between, there are polygenic traits with identifiable genetic causes. In monogenic diseases, a precise diagnosis is usually attainable; a clear genetic abnormality can be identified, and a biochemical explanation is often at hand. On the contrary, when we deal with psychological traits, we face difficulties even in attempting to define them (e.g., later we consider the cases of novelty seeking and partner choice). Furthermore, the underlying genetic modifications may be very subtle. Unfortunately, the moral capacity seems to be connected with strongly polygenic traits.

GENETIC EVIDENCES AND BEHAVIORAL PATHOLOGIES

Human Monogenic Diseases

It is not easy to find monogenic diseases that affect only the moral capacity. Indeed, there are many diseases affecting our capacity for reasoning, often grouped together under the heading "mental retardation." However, there are a few clinical conditions, that could be more pertinent to our topic, because they seem to concern behavior and the moral capacity. Let us consider a few of them.

GILLES DE LA TOURETTE SYNDROME (MIM 137580).[3] This syndrome has a very pleasant, aristocratic, name, and indeed the first description is due to Georges Gilles de la Tourette, a twenty-eight-year-old neurologist at the Hôpital de la Salpêtrière, who selected the life history of the Marquise of Dampierre, the cursing Marquise, as the prototypical example of the syndrome. Jean-Martin Charcot, the director of the Hôpital, renamed the convulsive tic syndrome in his honor. Tourette syndrome is a neurological disorder manifested particularly by motor and vocal tics and associated with behavioral abnormalities. Among these, coprolalia is the most famous, although rather infrequent (it is present in only about 8 percent of the cases; cf. Goldenberg et al. 1994). Another behavioral abnormality is a tendency to self-mutilation, which is present in 43 percent of cases (Van Woert et al. 1977). In spite of many efforts, the gene has not been identified yet, although it has been mapped in the 11q23 chromosomal region (Merette et al. 2000).

---

[3] MIM: Mendelian Inheritance of Man. It lists all the known hereditary diseases, identified with a code number. See http://www3.ncbi.nlm.nih.gov/omim/.

BRUNNER SYNDROME (MIM 309850). Another interesting case is represented by monoamine oxidase, an enzyme that catalyzes the oxidative deamination of biogenic amines throughout the body. This enzyme is critical in the neuronal metabolism of catecholamine transmitters. In 1993 Brunner and co-workers described a Dutch kindred in which all affected males showed characteristic behavioral abnormalities, in particular aggressive and sometimes violent behavior. Other types of impulsive behavior included attempts of rape and exhibitionism. In the eight affected males, the authors detected a nonsense mutation in the MAO-A gene (one of the two genes, MAO-A and MAO-B, which are grouped together in human X chromosome). Studies on mice revealed that an increase in aggressiveness co-varied with the inactivation of mice MAO-A genes. (It must be noted, however, that the inactivation of several other genes also co-varies with an increase in mice aggressiveness.) Unfortunately, after the case described by Brunner and co-workers, no other patients affected by MAO-A genetic mutations have been observed, although some pharmacological data support the hypothesis of the involvement of MAO-A in aggressiveness (Shih and Thompson 1999).

RETT SYNDROME (MIM 312750). People affected by this syndrome, after a normal development up to the age of seven to eighteen months, have a developmental stagnation, followed by rapid deterioration of high brain functions. Within 1.5 years, the deterioration process leads to severe dementia and autism. Interestingly, the gene responsible for this syndrome has been found (Amir et al. 1999): it codes for a CpG binding protein, which probably plays a role in the silencing of several genes.

SPEECH-LANGUAGE DISORDER IN THE KE FAMILY (MIM 602081). The KE family is affected by a rare and severe language and speech deficit, which segregates as an autosomal dominant trait in three generations (Fisher et al. 1998). It is one of the most selective hereditary language abnormalities, although the pathogenesis of the defect is not clear. Some authors suggest that it is a specific language defect, whereas others believe that it is a defect in the coordination of the ultra-rapid movements that are necessary to produce comprehensible speech (Watkins et al. 1999). The responsible gene has recently been found (Lai et al. 2001), and the mutation of the KE family has been identified. In addition, a different abnormality in the same gene has been detected in another patient, unrelated to the KE family, but manifesting the same symptoms. The involved gene codes for a protein belonging to a family that includes some members known to play a role in fetal development.[4]

Monogenic Abnormalities in Mice

For obvious ethical reasons, we cannot perform experiments on humans. Nevertheless, we can experiment on mice and thus investigate the effects of the modification of their genes. This technique, pioneered by Mario Capecchi (1989), is based on the homologous recombination of embryonic stem cells and allows us to inactivate any kind of known gene, including the genes potentially involved in behavior control that we are interested in. In what follows, we mention just a few of these experiments. It is contentious, however, whether their results can be straightforwardly extended to humans, because it is widely accepted that the most complex human kinds of behavior do not have precise counterparts in mice (on this topic, see Bucan and Abel 2002; Fasolo, Chapter 4 in this volume).

GENES AND ANXIETY. Many genes have been involved in the pathogenesis of anxiety in experimental models. In 1998 three papers reported that the inactivation of the 1A serotonin receptor causes an increase of anxious behaviors in mice (Parks et al. 1998; Ramboz et al. 1998; Heisler et al. 1998). Recently, it has been shown that hippocampal and cortex neurons expressing this receptor are responsible for this increase (Gross et al. 2002). Crestani and co-workers in 1999 reported that mice heterozygous for a receptor for GABA (gamma-aminobutyric acid), the major inhibitory neurotransmitter in the mammalian brain, display anxious behavior. Still in 1999, an increase in anxious behavior was also reported by Karolyi and co-workers in mice knockout for the CRH-BP (corticotropin-releasing hormone-binding protein) gene and in mice in which the alpha4 nicotinic receptor had been disrupted (Labarca et al. 2001). Likewise, mice in which the gene for the opioid peptide enkephalin was inactivated displayed exaggerated responses to situations evoking fear or anxiety (Ragnauth et al. 2001). The same phenotype was shown by mice with knockout of the delta opioid receptor, the receptor through which only enkephalins can act effectively (Filliol et al. 2000). On the contrary, mice in which the glucocorticoid receptor gene had been specifically inactivated in the nervous system had decreased levels of anxiety (Tronche et al. 1999).

---

[4] The evolution of this gene has been investigated in chimpanzees and other species, and the possibility of its contribution to language acquisition has been raised; see note 15.

AGGRESSIVENESS AND NITRIC OXIDE (NO). Mice knockout for neuronal NO synthase (nNOS) shows an extreme aggressiveness (Nelson et al. 1995), which is not due to a defect in embryonic development, because pharmacological inhibition of this enzyme mimics the same phenotype. Therefore, it would seem a relatively pure behavioral abnormality.

MONOGAMY AND VASOPRESSIN. During the past few years there were several reports on two very similar rodent species, the prairie voles and the mountain voles, only the first of which is monogamous. These studies elucidate the role of a hormone, vasopressin, in the genesis of some kinds of mating behavior (Carter and Getz 1993). Starting from these studies, transgenic mice were produced in which the prairie vole expression pattern of a vasopressin receptor had been reconstituted. These transgenic mice show an increase in their affiliative behavior. This study, if confirmed, could be interesting because this behavior modification is very selective and no other abnormality is evident in these mice (Young et al. 1999).[5]

SOCIAL MEMORY AND OXYTOCIN. Mice in which the oxytocin gene was interrupted show a defect in social memory (Ferguson et al. 2000), operatively defined as a reliable decrease in olfactory investigation in repeated or prolonged encounters with others of their breed. Mutant males do not have the ability to recognize individuals that they have previously encountered. The ability can be restored by a single injection of oxytocin. Hence, the defect is not due to a neurological abnormality occurring during embryo development.

MATERNAL BEHAVIOR AND ONCOGENIS. A somewhat unexpected result was obtained from the investigation of the role of oncogenes in murine development. Surprisingly, mice in which the c-FosB gene was interrupted showed a defect in maternal behavior (Brown et al. 1996). They were profoundly deficient in their ability to nurture young animals but were normal with respect to other cognitive and sensory functions. Therefore the deficit looks quite specific.

GROOMING AND HOMEOBOX. In another recent study, Green and Capecchi (2002) engineered mice with mutations in Hoxb8 gene, a homeobox gene. These mice exhibited a selective behavioral abnormality, excessive grooming, as they groomed themselves to the point of self-mutilation. In primates,

---

[5] This topic has been recently reviewed in Young and Wang 2004.

grooming is a social activity aimed at reinforcing cooperative links. Interestingly, a pathology similar to that described in these mice has also been described in humans (Graybiel and Saka 2002).

ATYPICAL SEXUAL BEHAVIOR. Mice with abnormalities in sexual or aggressive behavior have been described, but recently a striking sexual behavioral phenotype has been obtained following the inactivation of the transient receptor potential-2 (TRP2) gene. Stowers and co-workers inactivated this gene, which is specifically expressed in an olfactive structure, the vomeronasal organ (VNO), which responds to pheromones, molecules that have been involved in social interactions, including sexual discriminations of conspecifics (Stowers et al. 2002; cf. also the comment by Keverne 2002). Mice deficient in TRP2 expression failed to display aggressive behavior against males and initiated sexual activity with both males and females indiscriminately. Because the product of the TRP2 gene is a pheromone receptor, the pathogenetic mechanism would be obvious, since it can be easily hypothesized that the $TRP2^{-/-}$ mouse is unable to perceive a specific molecule (whose exact nature is still unknown) signaling gender specificity. According to the authors, mating could be the default behavior, which is inhibited by a male-specific pheromone. If the mice detect the presence of this molecule, the aggressive program is triggered, and if not, the default program is maintained, leading to mating.[6]

SMART MICE. Usually, gene targeting in mice creates pathological phenotypes. However, there is a report of mice showing better performances after overexpression of the beta type receptor for NMDA (N-methyl-D-aspartate) in their brains. These mice exhibit superior learning and memory abilities in various behavioral tasks. The authors conclude that genetic enhancement of mental and cognitive attributes such as intelligence and memory in mammals is feasible (Tang et al. 1999). On the other hand, it is quite clear that the same

---

[6] Studies evaluating aggressiveness or mating choices are made according to the resident-intruder protocol by introducing a male or female mouse (intruder) in a cage where an isolated mouse had established territory (resident). Normally, the male intruder elicits aggressive behavior from the resident (triggered by pheromones in the intruder's urine), while the female intruder does not (in this case the resident engages in sexual behaviors). In Stowers's study, the resident is a $TRP^{-/-}$ male (in which the TRP2 gene has been inactivated); it showed no sign of aggression. On the contrary, the $TRP^{-/-}$ resident male approached the intruder and engaged in sexual behavior. Indeed, he mounted both males and females with the same frequency when they were introduced together in the cage. For this reason the authors suggest that mating, not fighting, is the default behavior. As it has been remarked (Beckman 2002), these studies suggest that love is fundamentally more important than war.

effects can be achieved by an increase in environmental stimuli (van Praag et al. 2000).

Human Polygenic Diseases

Let us consider now the polygenic genetic diseases leading to behaviors that are usually considered abnormal. The notion of normality involved here is a tricky one. Cases of monogenic diseases are simpler, because we know what "normal" means: indeed, we can identify, at least in principle, the "normal" nucleotide sequence of a given gene and the "abnormal" variations from it. "Normal" and "abnormal" kinds of behavior will then be the products of "normal" and "abnormal" nucleotide sequences respectively. In polygenic disease, on the contrary, more than one gene is involved; and each one can contribute to behavior in many different ways (each one might not express its protein, or express it at a degree different from usual, or it might express an unusual protein, etc.). Moreover, in many cases, an extremely important role is played by the genetic and extragenic environments and their interactions with the genes involved. What counts as a genetic "normality" or "abnormality," then, varies according to the combination of all these factors.

We consider some of the most debated cases in the past ten years – homosexuality, partner choice, and novelty seeking – and also add a recent case of genetic alterations linked to panic disorders.

HOMOSEXUALITY. Homosexuality is a condition that, in some aspects, involves a moral behavior and therefore is a useful example. Some studies reported the existence of anatomical differences between the brains of normal and homosexual males (e.g., LeVay 1991), but these differences do not constitute conclusive evidence, because they might be an effect and not a cause of homosexuality. In 1993 a report published in *Science* gave rise to extensive debates. In this article, Hamer and co-workers established a linkage between male homosexuality and the chromosomal region Xq28.[7] This article had great impact, and two years later the same authors added new data supporting their thesis (Hu et al. 1995). However, in 1999, a large Canadian study challenged Hamer's results.[8] Although at the moment it is not

[7] Hamer et al. 1993. See also the very optimistic accompanying editorial ("Evidence for Homosexuality Gene"), in the same issue of *Science* (261: 291).

[8] Rice et al. 1999. See also the discussion in Wickelgren 1999 and the correspondence in "Genetics and Male Sexual Orientation," *Science* 285 (1999): 803. One of the authors of Rice et al. 1999, the statistical genetics expert Neil Risch, had already raised many doubts in 1993. See correspondence to the 1993 paper by Rice in *Science* 262 (1993): 2063–2065.

completely clear who is right, the initial enthusiasm surrounding this topic has vanished.

NOVELTY SEEKING. More or less the same fate was shared by studies linking a behavioral trait to a gene coding for a dopamine receptor. In 1996 two studies, one again by Hamer, claimed that carriers of a polymorphism in this receptor are more likely to be classified as novelty seekers, that is, people who like challenges (Benjamin et al. 1996; Ebstein et al. 1996; Ebstein et al. 1997). However, further investigations failed to confirm this association (Malhotra et al. 1996; Jonsson et al. 1997; Gelernter et al. 1997; Pogue-Geile et al. 1998). Also in this case, enthusiasm has since waned.

PARTNER CHOICE. As mentioned previously,[9] chemical cues perceived through olfaction can be relevant for social behavior. It has long been known that in rodents the olfactory system plays a fundamental role, whereas in humans it has been largely taken over by alternative sensory modalities. However, olfaction could still mediate some kinds of sexual behavior, such as the choice of a partner. This possibility has recently been supported by a study reporting that women can detect olfactorially differences of an HLA allele, discriminating males with different genotypes at the histocompatibility complex.[10] Apparently, women's choices are based on HLA alleles inherited from the fathers but not from the mothers.[11] Obviously, these results need to

---

[9] See the discussion of atypical sexual behavior in the previous section. With reference to this point, it is worth recalling a recent article (Krieger and Ross 2002; see also the comments by Holden 2001 and Crozier 2002), which, for the first time, reports a straightforward correlation between a DNA polymorphism and a complex social behavior in a fire ant species, *Solenopsis invicta*. Within this species, some colonies have a single queen (monogyne social form), whereas others have multiple queens (polygyne). Very elegantly, the authors were able to demonstrate that monogyne queens are homozygous for an allele at the Gp-9 locus (BB), whereas polygyne queens are heterozygous (Bb). The Gp-9 locus codes for a pheromone-binding protein. In this case, genetic findings are very cogent, while the cloning of the involved gene gives us a physiological mechanism to explain the phenomenon: pheromones are key components in chemical recognition among the members of the same species ("conspecies").

[10] Jacob et al. 2002. The histocompatibility system has long been involved in social recognition in rodents, see, for example, Singh et al. 1987. In this article, the authors claim that mice can discriminate the odors of animals differing from a single histocompatibility allele. Another recent article has identified a specific protein class, the major urinary protein, as the mediator of identity signals in the context of territorial mouse behavior. See Hurst et al. 2001.

[11] The histocompatibility system is an extremely polymorphic set of gene products that is also responsible for transplant rejection. Because transplants did not occur throughout evolution, it is obvious to think that histocompatibility systems evolved for other reasons, and indeed they are now known to play a fundamental role in the immune response. However, it is not unlikely that they play additional roles in social recognition. It would be another example of pleiotropism.

be repeated and confirmed by other data before their significance in social and evolutionary terms can be taken for granted.

PANIC DISORDERS. A very interesting result has been reported by Xavier Estivill and his associates. They found that a genomic duplication of a portion of the 15q chromosome, apparently inherited in a non-Mendelian way, is strongly associated with phobic disorders (Gratacos et al. 2001). They took advantage of the occasional association between a physical characteristic, joint laxity, and phobia itself. An association of this kind is very useful in the investigation of psychiatric diseases, because it may help psychiatric diagnoses that are otherwise very difficult. Classic examples of these difficulties could be depression and mental retardation. Even in people with a genetically depressed ancestor, depression could derive from environmental causes. Similarly, even in people with retarded parents, a mental deficit could completely depend on the poverty of stimulations, which presumably any genetically normal child would experience in a family of retarded people.

The association between panic disorders and the duplication described by Estivill and co-workers is very strong, because about the 80 percent of the original pedigrees with panic disorders bore this genomic abnormality. This duplication is present in 6 percent of the general population but in 95 percent of the phobic patients, unrelated to the original pedigrees. Hence, it is likely that other genes contribute to this disorder as well. The significance of these results, however, has recently been questioned (Tabiner et al. 2003; Henrichsen et al. 2004; Zhu et al. 2004).

PHILOSOPHICAL REMARKS ON GENETIC EVIDENCE

The genetic results reviewed here suggest some interesting considerations about the human moral capacity and its enabling conditions.

First of all, it should be noted that a large part of the research on the relationship between genes and moral capacity focuses on sexual or feeding-related aspects. Both sexual and feeding behaviors often involve aspects of moral judgment. It is very likely that these kinds of behavior, which evolved over hundreds of millions of years, depend to a large extent on genes that can also be found in many other species. But these kinds of behavior do not represent the complexity of human actions completely, and it would be inappropriate to generalize on the basis of such behaviors. One must be aware of the risk of committing the *fallacy of the false generalization.*

Likewise, we must avoid *confusions arising from the use of terms*, which can have very different meanings in ethology and in everyday human situations. Take as an example the term "reciprocal altruism." When applied to humans in ordinary language, the term "altruism" has really almost nothing to do with the same term, as used in current ethological terminology. Moreover, even though it is persuasively suggested that reciprocal altruism has a genetic basis, still we do not know much about its putative molecular basis, which is what really matters when discussing its biological roots (cf. Rosenberg, Chapter 10 in this volume).

Our discussion raises a second consideration. Generally speaking, in humans and – partially – in mice, the inactivation of a single gene may influence a kind of behavior. What usually results is a *deviant* modification of a specific behavior, or the disruption of the normal balance of the organism. This does not seem a good enough ground to claim that that gene, in its nondeviant form, has to do with behavior or, even worse, that behavior is genetically determined. With the loss of legs, one cannot walk anymore. But this is not to say that one's walking was previously *determined* by its legs.

Moreover, the function of the genes involved, when at least partially known, is completely aspecific and usually concerns very basic biochemical and cellular pathways. Therefore, so far, investigations of monogenic diseases have not offered thus far a decisive contribution to behavioral genetics, even if, as shown in the cases of Gilles de la Tourette syndrome, Brunner syndrome, Rett syndrome, and mutant mice, the change in only one gene influences, even drastically, moral capacity and behavior (respectively, we have coprolalia and self-mutilation, aggressiveness and exhibitionism, dementia, anxiety, loss of monogamy and social memory, and so on).

In other terms, it must be recognized that behavioral genetics is still in its infancy, in the sense that even if we know something about the behavioral effects of some single gene mutations, almost nobody has attempted to establish extensive pedigrees for complex behavioral traits.[12] This might depend on the fact that such pedigrees do not exist (and, thus, these traits do not run in families because they do not have any genetic basis), or that nobody believes they have a genetic basis (therefore nobody wants to waste time searching for them), or that reliable tests to measure these traits are not available (e.g., it is impossible to establish pedigrees because it is empirically impossible to identify and follow the relevant trait through generations).

---

[12] The description of a large family (pedigree) with several members manifesting a given phenotype is the starting point for the identification of the genes responsible for that phenotype.

With regard to the polygenic transmission of behavioral characters, as it was supposed in the case of homosexuality, novelty seeking, and partner choice, the issue is extremely difficult also for technical reasons. Unless some fortunate circumstance occurs, like that which led to the identification of the chromosome 15q duplication involvement in panic disorders, which we considered previously, current genetic techniques are not yet ready to tackle the problem efficiently. Will it be possible to establish whether genes influence, at least partially, the moral capacity? Probably yes. Surely we need more powerful instruments, a large sequencing output, and new bioinformatic tools. If we consider the problem carefully, we are currently working relatively randomly, trying to correlate small DNA segments with behavioral traits. It is likely that an answer could be obtained, one way or another, if sequencing techniques improved their output by 100 or even 1,000 times.

Today, the chimpanzee genome is available for analysis and comparison with the human one. It is commonly held that the nucleotide differences between humans and chimps are around 1 or 2 percent, and it would be extremely interesting to know what that 1 or 2 percent is. It is possible that the comparison between *Homo* and *Pan* DNA sequences will provide interesting hints about the basis of the differences, including behavioral differences, between our species and other apes. The point is not establishing that several differences between man and chimp have genetic bases. This is already beyond any reasonable doubt. The interesting thing, perhaps only a dream, would be to understand whether and to what degree our moral capacity is conditioned by genes, and which brain structures make it possible for individuals to take different behavioral paths after having morally assessed the available possibilities. That is, it would be important to know whether and to what extent we can freely choose our morally relevant actions.

What can be certainly asserted at this point in the history of science, though, seems to be that there are not enough grounds to claim that our moral capacity is fully determined by our genes. Indeed, even in some monogenic diseases – which are the most likely to be deterministically explained – genes cannot completely determine individual moral capacity. The course of these diseases can be affected by the surrounding environment, including man. It is enough to cite phenylketonuria (MIM 261600),[13] the stigmata of which can be completely reversed with an early diagnosis and a strict diet. In the case of polygenic diseases, things are even vaguer; indeed, more than polygenic,

---

[13] http://www3.ncbi.nlm.nih.gov/htbin-post/Omim/dispmim? 261600.

they are multifactorial, in the sense that their clinical course depends largely on external circumstances. Therefore an individual with all the wrong genes may never get sick at all.

Many researchers believe that genes provide only a set of *structures* whose manifestation is largely undetermined. There are many such cases. Let us analyze hand joints; there is no doubt that they are the result of a genetic pattern. But whether they will be used to grab a knife, play the violin, or shoot the ball into a basket is not – surely – written in the genes. The idea that genes and structures can be pleiotropic is completely plausible.[14] There is no doubt that in a few centuries we will have much clearer ideas on this topic – and many surprises.

The language example is emblematic. If it is true that some changes (in a broad sense) in primate genes, occurring over a period of 5–6 million years, have determined the capacity to speak in a primate species, it is also true that the specific language spoken by an individual is largely undetermined by his genes. Language can be thought of as a behavioral trait, and by comparing the genetics of man and apes we will be able to identify the genetic changes that led to this.[15] By comparing morphological and neurophysiological data we will probably be able to elucidate the structures, due to genetic changes, that play a role in language abilities. But in spite of this, presumably we will not be able to predict what language will be spoken by a single person.

Now, there is no reason to think that a given behavior can be determined any more than a given language is. On the contrary, there is a relatively large consensus on the idea that behaviors can be even more underdetermined than languages.

The conclusion that we feel sure to suggest is that even if we know, by studying monogenic and polygenic diseases, that our genes, in particular their deviant forms, influence our moral capacity by acting on its enabling conditions, there is not enough scientific ground yet to state to what degree these influences occur. As suggested at the beginning of this chapter, we

---

[14] Ayala 2000. The secretion, in urines, of compounds that can trigger social behavior in rodents is a clear example of pleiotropy. It is obvious that the excretory renal system did not develop in order to signal our existence to others, but some of its features have apparently been "co-opted" for novel unpredictable aims.

[15] A first step in this direction has recently been reported; see Enard et al. 2002. In this article, the FOXP2 gene, which is responsible for the language disorder just reported, was sequenced in chimpanzee, gorilla, orangutan, rhesus macaque, and mouse and compared with the human sequence. Compared with primate DNA, the human FOXP2 contains two amino acid changes, which could have played a role in the acquisition of brain structures involved with the ability to speak.

should remember Locke's lesson and limit ourselves to the little we know, without any overbearing philosophically useless generalization. We believe that, concerning the genetic influences on moral capacity, we cannot, so far, make any strong general claims. However, it is possible that when very large-scale sequencing and novel approaches to genome analysis become available, we will be able to find empirical data relevant for a more precise philosophical assessment of the entire matter.

REFERENCES

Amir, R. E., Van den Veyver, I. B., Wan, M., Tran, C. Q., Francke, U., and Zoghbi, H. Y. 1999. Rett Syndrome Is Caused by Mutations in X-linked MECP2, Encoding Methyl-CpG-binding Protein 2. *Nat. Genet.* 23: 185–188.

Ayala, F. J. 2000. From the Myth of Eden to a New Garden: Genetics and Ethical Responsibility. In *Life Sciences and the New Humanism*, 25–37. Bologna: Fondazione Marino Golinelli.

Beckman, M. 2002. When in Doubt, Mice Mate Rather Than Fight. *Science* 295: 782.

Benjamin, J., Li, L., Patterson, C., Greenberg, B. D., Murphy, D. L., and Hamer, D. H. 1996. Population and Familial Association between the D4-dopamine Receptor Gene and Measures of Novelty Seeking. *Nat. Genet.* 12: 81–84.

Berkowitz, A. 1996. Our Genes, Ourselves? *BioScience* 46: 42–51.

Brown, J. R., Ye, H., Bronson, R. T., Dikkes, P., and Greenberg, M. E. 1996. A Defect in Nurturing in Mice Lacking the Immediate Early Gene fos B. *Cell* 86: 297–309.

Brunner, H. G., Nelen, M., Breakefield, X. O., Ropers, H. H., and van Oost, B. A. 1993. Abnormal Behavior Associated with a Point Mutation in the Structural Gene for Monoamine Oxidase A. *Science* 262: 578–580.

Bucan, M., and Abel, T. 2002. The Mouse: Genetics Meets Behavior. *Nature Rev. Genet.* 3: 114–123.

Capecchi, M. R. 1989. Altering the Genome by Homologous Recombination. *Science* 244: 1288–1292.

Carter, S. C., and Getz, L. L. 1993. Monogamy and the Prairie Vole. *Scient. Am.* June: 100–106.

Crestani, F., Lorez, M., Baer, K., Essrich, C., Benke, D., Laurent, J. P., Belzung, C., Fritschy, J. M., Luscher, B., and Mohler, H. 1999. Decreased GABAA-Receptor Clustering Results in Enhanced Anxiety and a Bias for Threat Cues. *Nat. Neurosci.* 2: 833–839.

Crozier, R. H. 2002. Pheromones and the Single Queen. *Nat. Genet.* 30: 4–5.

de Belle, J. S. 2002. Unifying the Genetics of Behavior. *Nat. Genet.* 31: 1–2.

Ebstein, R. P., Nemanov, L., Klotz, I., Gritsenko, I., and Belmaker, R. H. 1997. Additional Evidence for an Association between the Dopamine D4 Receptor (D4DR) exon III Repeat Polymorphism and the Human Personality Trait of Novelty Seeking. *Mol. Psychiatry* 2: 472–477.

Ebstein, R. P., Novick, O., Umansky, R., Priel, B., Osher, Y., Blaine, D., Bennett, E. R., Nemanov, L., Katz, M., and Belmaker, R. H. 1996. Dopamine D4 (D4DR) exon III Polymorphism Associated with the Human Personality Trait of Novelty Seeking. *Nat. Genet.* 12: 78–80.

Enard, W., Przeworski, M., Fisher, S. E., Lai, C. S. L., Wiebe, V., Kitano, T., Monaco, A., and Pääbo, S. 2002. Molecular Evolution of FOXP2, a Gene Involved in Speech and Language. *Nature* 418: 869–872.

Ferguson, J. N., Young, L. J., Hearn, E. F., Matzuk, M. M., Insel, T. R., and Winslow, J. T. 2000. Social Amnesia in Mice Lacking the Oxytocin Gene. *Nat. Genet.* 25: 284–288.

Filliol, D., Ghozland, S., Chluba, J., Martin, M., Matthes, H. W., Simonin, F., Befort, K., Gaveriaux-Ruff, C., Dierich, A., LeMeur, M., Valverde, O., Maldonado, R., and Kieffer, B. L. 2000. Mice Deficient for Delta- and Mu-Opioid Receptors Exhibit Opposing Alterations of Emotional Responses. *Nat. Genet.* 25: 195–200.

Fisher, S. E., Vargha-Khadem, F., Watkins, K. E., Monaco, A. P., and Pembrey, M. E. 1998. Localisation of a Gene Implicated in a Severe Speech and Language Disorder. *Nat. Genet.* 18: 168–170.

Garcia, J., McGowan, B. K., and Green, K. F. 1972. Biological Constraints on Learning. In *Biological Boundaries of Learning*, ed. M. E. P. Seligman and J. L. Hager, 21–43. New York: Appleton Century Crofts.

Gelernter, J., Kranzler, H., Coccaro, E., Siever, L., New, A., and Mulgrew, C. L. 1997. D4 Dopamine-Receptor (DRD4) Alleles and Novelty Seeking in Substance-Dependent, Personality-Disorder and Control Subjects. *Am. J. Hum. Genet.* 61: 1144–1152.

Gilbert, W. 1992. Vision of the Grail. In *The Code of Codes: Scientific and Social Issue in the Human Genome Project*, ed. D. J. Kelves and L. Hood, 83–97. Cambridge, Mass.: Harvard University Press.

Goldenberg, J. N., Brown, S. B., and Weiner, W. J. 1994. Coprolalia in Younger Patients with Gilles de la Tourette Syndrome. *Mov. Disord.* 9: 622–625.

Gratacos, M., Nadal, M., Martin-Santos, R., Pujana, M. A., Gago, J., Peral, B., Armengol, L., Ponsa, I., Miro, R., Bulbena, A., and Estivill, X. 2001. A Polymorphic Genomic Duplication on Human Chromosome 15 Is a Susceptibility Factor for Panic and Phobic Disorders. *Cell* 106: 367–379.

Graybiel, A. M., and Saka, E. 2002. A Genetic Basis for Obsessive Grooming. *Neuron.* 33: 1–2.

Green, J. M., and Capecchi, M. R. 2002. Hoxb8 Is Required for Normal Grooming Behavior in Mice. *Neuron.* 33: 23–34.

Gross, C., Zhuang, X., Stark, K., Ramboz, S., Oosting, R., Kirby, L., Santarelli, L., Beck, S., and Hen, R. 2002. Serotonin1A Receptor Acts during Development to Establish Normal Anxiety-Like Behavior in the Adult. *Nature* 416: 396–400.

Hamer, D. H., Hu, S., Magnuson, V. L., Hu, N., and Pattatucci, A. M. 1993. A Linkage between DNA Markers on the X Chromosome and Male Sexual Orientation. *Science* 261: 321–327.

Harlow, H. F., Dodsworth, R. O., and Harlow, M. K. 1965. Total Isolation in Monkeys. *Proc. Natl. Acad. Sci. USA* 54: 90–97.

Heisler, L. K., Chu, H. M., Brennan, T. J., Danao, J. A., Bajwa, P., Parsons, L. H., and Tecott, L. H. 1998. Elevated Anxiety and Antidepressant-Like Responses in Serotonin 5-HT1A Receptor Mutant Mice. *Proc. Natl. Acad. Sci. USA* 95: 15049–15054.

Henrichsen, C. N., Delorme, R., Boucherie, M., Marelli, D., Baud, P., Bellivier, F., Courtet, P., Chabane, N., Henry, C., Leboyer, M., Malafosse, A., Antonarakis, S. E., and Dahoun, S. 2004. No Association between DUP25 and Anxiety Disorders. *Am. J. Med. Genet.* 128B: 80–83.

Holden, C. 2001. Single Gene Dictates Ant Society. *Science* 294: 1434.

Hu, S., Pattatucci, A. M., Patterson, C., Li, L., Fulker, D. W., Cherny, S. S., Kruglyak, L., and Hamer, D. H. 1995. Linkage between Sexual Orientation and Chromosome Xq28 in Males but Not in Females. *Nat. Genet.* 11: 248–256.

Hurst, J. L., Payne, C. E., Nevison, C. M., Marie, A. D., Humphries, R. E., Robertson, D. H., Cavaggioni, A., and Beynon, R. J. 2001. Individual Recognition in Mice Mediated by Major Urinary Proteins. *Nature* 414: 631–644.

Jacob, S., McClintock, M. K., Zelano, B., and Ober, C. 2002. Paternally Inherited HLA Alleles Are Associated with Women's Choice of Male Odor. *Nat. Genet.* 30: 175–179.

Jaroff, L. 1989. The Gene Hunt. *Time*, 20 March: 62–67.

Jonsson, E. G., Nothen, M. M., Gustavsson, J. P., Neidt, H., Brene, S., Tylec, A., Propping, P., and Sedvall, G. C. 1997. Lack of Evidence for Allelic Association between Personality Traits and the Dopamine D4 Receptor Gene Polymorphisms. *Am. J. Psychiatry* 154: 697–699.

Karolyi, I. J., Burrows, H. L., Ramesh, T. M., Nakajima, M., Lesh. J. S., Seong, E., Camper, S. A., and Seasholtz, A. F. 1999. Altered Anxiety and Weight Gain in Corticotropin-Releasing Hormone-Binding Protein-Deficient Mice. *Proc. Natl. Acad. Sci. USA* 96: 11595–11600.

Keverne, E. B. 2002. Pheromones, Vomeronasal Function and Gender-Specific Behavior. *Cell* 108: 735–738.

Kitcher, P. 2001. Battling the Undead: How (and How Not) to Resist Genetic Determinism. In *Thinking about Evolution: Historical, Philosophical, and Political Perspectives*, ed. R. S. Singh, C. B. Krimbas, and J. Beattie, 396–414. Cambridge: Cambridge University Press.

Krieger, M. J., and Ross, K. G. 2002. Identification of a Major Gene Regulating Complex Social Behavior. *Science* 295: 328–332.

Labarca, C., Schwarz, J., Deshpande, P., Schwarz, S., Nowak, M. W., Fonck, C., Nashmi, R., Kofuji, P., Dang, H., Shi, W., Fidan, M., Khakh, B. S., Chen, Z., Bowers, B. J., Boulter, J., Wehner, J. M., and Lester, H. A. 2001. Point Mutant Mice with Hypersensitive Alpha 4 Nicotinic Receptors Show Dopaminergic Deficits and Increased Anxiety. *Proc. Natl. Acad. Sci. USA* 98: 2786–2791.

Lai, C. S., Fisher, S. E., Hurst, J. A., Vargha-Khadem, F., and Monaco, A. P. 2001. A Forkhead-Domain Gene Is Mutated in a Severe Speech and Language Disorder. *Nature* 413: 519–523.

LeVay, S. 1991. A Difference in Hypothalamic Structures between Heterosexual and Homosexual Men. *Science* 253: 1034–1037.

Lewontin, R. C. 1991. *Biology as Ideology*. New York: Harper Perennial.

Lewontin, R. C., S. Rose, and Kamin, L. J. 1984. *Not in Our Genes: Biology, Ideology and Human Nature*. New York: Pantheon.

Locke, J. 1690. *An Essay Concerning Human Understanding*. Oxford: Clarendon Press, 1975.

Malhotra, A. K., Virkkunen, M., Rooney, W., Eggert, M., Linnoila, M., and Goldman, D. 1996. The Association between the Dopamine D4 Receptor (D4DR) 16 Amino Acid Repeat and Novelty Seeking. *Mol. Psychiatry* 1: 388–391.

Merette, C., Brassard, A., Potvin, A., Bouvier, H., Rousseau, F., Emond, C., Bissonnette, L., Roy, M.-A., Maziade, M., Ott, J., and Caron, C. 2000. Significant Linkage for

Tourette Syndrome in a Large French Canadian Family. *Am. J. Hum. Genet.* 67: 1008–1013.

Nelson, R. J., Demas, G. E., Huang, P. L., Fishman, M. C., Dawson, V. L., Dawson, T. M., and Snyder, S. H. 1995. Behavioural Abnormalities in Male Mice Lacking Neuronal Nitric Oxide Synthase. *Nature* 378: 383–386.

Oyama, S. 1985. *The Ontogeny of Information: Developmental Systems and Evolution.* Cambridge: Cambridge University Press.

Parks, C. L., Robinson, P. S., Sibille, E., Shenk, T., and Toth, M. 1998. Increased Anxiety of Mice Lacking the Serotonin1A Receptor. *Proc. Natl. Acad. Sci. USA* 95: 10734–10739.

Pogue-Geile, M., Ferrell, R., Deka, R., Debski, T., and Manuck, S. 1998. Human Novelty-Seeking Personality Traits and Dopamine D4 Receptor Polymorphisms: A Twin and Genetic Association Study. *Am. J. Med. Genet.* 81: 44–48.

Ragnauth, A., Schuller, A., Morgan, M., Chan, J., Ogawa, S., Pintar, J., Bodnar, R. J., and Pfaff, D. W. 2001. Female Preproenkephalin-knockout Mice Display Altered Emotional Responses. *Proc. Natl. Acad. Sci. USA* 98: 1958–1963.

Ramboz, S., Oosting, R., Amara, D. A., Kung, H. F., Blier, P., Mendelsohn, M., Mann, J. J., Brunner, D., and Hen, R. 1998. Serotonin Receptor 1A Knockout: An Animal Model of Anxiety-Related Disorder. *Proc. Natl. Acad. Sci. USA* 95: 14476–14481.

Rice, G., Anderson, C., Risch, N., and Ebers, G. 1999. Male Homosexuality: Absence of Linkage to Microsatellite Markers at Xq28. *Science* 284: 665–667.

Risch, N. J. 2000. Searching for Genetic Determinants in the New Millennium. *Nature* 405: 847–856.

Shih, J. C., and Thompson, R. F. 1999. Monoamine Oxidase in Neuropsychiatry and Behavior. *Am. J. Hum. Genet.* 65: 593–598.

Singh, P. B., Brown, R. E., and Roser, B. 1987. MHC Antigens in Urine as Olfactory Recognition Cues. *Nature* 327: 161–164.

Stowers, L., Holy, T. E., Meister, M., Dulac, C., and Koentges, G. 2002. Loss of Sex Discrimination and Male-Male Aggression in Mice Deficient for TRP2. *Science* 295: 1493–1500.

Tabiner, M., Youings, S., Dennis, N., Baldwin, D., Buis, C., Mayers, A., Jacobs, P. A., and Crolla, J. A. 2003. Failure to Find DUP25 in Patients with Anxiety Disorders, in Control Individuals, or in Previously Reported Positive Control Cell Lines. *Am. J. Hum. Genet.* 72: 535–538.

Tang, Y. P., Shimizu, E., Dube, G. R., Rampon, C., Kerchner, G. A., Zhuo, M., Liu, G., and Tsien, J. Z. 1999. Genetic Enhancement of Learning and Memory in Mice. *Nature* 401: 63–69.

Tronche, F., Kellendonk, C., Kretz, O., Gass, P., Anlag, K., Orban, P. C., Bock, R., Klein, R., and Schutz, G. 1999. Disruption of the Glucocorticoid Receptor Gene in the Nervous System Results in Reduced Anxiety. *Nat. Genet.* 23: 99–103.

van Praag, H., Kempermann, G., and Gage, F. H. 2000. Neural Consequences of Environmental Enrichment. *Nat. Rev. Neurosci.* 1: 191–198.

Van Woert, M. H., Yip, L. C., and Balis, M. E. 1977. Purine Phosphoribosyltransferase in Gilles de la Tourette Syndrome. *New Engl. J. Med.* 296: 210–212.

Vezzoni, P. 2000. *Intersezioni. Questioni biologiche di rilevanza filosofica.* Milan: McGraw-Hill.

Watkins, K. E., Gadian, D. G., and Vargha-Khadem, F. 1999. Functional and Structural Brain Abnormalities Associated with a Genetic Disorder of Speech and Language. *Am. J. Hum. Genet.* 65: 1215–1221.

Wickelgren, I. 1999. Discovery of Gay Gene Questioned. *Science* 284: 571.

Young, L. J., Nilsen, R., Waymire, K. G., MacGregor, G. R., and Insel, T. R. 1999. Increased Affiliative Response to Vasopressin in Mice Expressing the V1a Receptor from a Monogamous Vole. *Nature* 400: 766–768.

Young, L. J., and Wang, Z. 2004. The Neurobiology of Pair Bonding. *Nat. Neurosci.* 7: 1048–1054.

Zhu, G., Bartsch, O., Skrypnyk, C., Rotondo, A., Akhtar, L. A., Harris, C., Virkkunen, M., Cassano, G., and Goldman, D. 2004. Failure to Detect DUP25 in Lymphoblastoid Cells Derived from Patients with Panic Disorder and Control Individuals Representing European and American Populations. *Eur. J. Hum. Genet.* 12: 505–508.

6

Evolutionary Psychopharmacology, Mental Disorders, and Ethical Behavior

STEFANO CANALI, GABRIELE DE ANNA, AND LUCA PANI

The concept of pathology has a built-in normative character. An individual (or a state) is pathological if and only if it is not normal, that is, if it fails to be as an individual (or a state) of that sort *ought to be*. In other words, a pathological individual (or state) is an individual (or state) that does not conform to the standards to which all the individuals (or states) falling under its very sortal concept *must* conform.

When applied to mental disorders, this view entails that the concept of psychopathology has a normative character. A human being is psychopathic if and only if he fails to have the mental capacities that are normal for humans, that is, the mental capacities that humans *ought to have*. Human specific mental capacities include emotional responses to the environment, as well as perceptual and cognitive abilities. Let us call this the "received view of psychopathology."

The received view has two main implications concerning psychiatry. The first implication is that psychiatry is ethically relevant, in two respects. First, the results of psychopathology may help to define what is normal for humans, and thus psychiatry may help to determine ethical norms. Second, the psychiatric treatment of mental disorders rests on various assumptions concerning what is normal for humans, and thus psychiatry has profound ethical bearings.

The second implication of the received view of psychopathology is what we could call the "universal treatment thesis." This is the idea that each mental disorder must have one perfectly appropriate cure, which scientists have to work out and clinical psychiatrists have to apply to individual patients. Naturally, the universal treatment thesis does not rest on the received view alone, but it is also grounded on at least two other complementary assumptions. First, humans (which, in this case, constitute the set of normal individuals setting the normative standards) are sufficiently similar to each other for it to be the case that two individuals who have equally nonnormal mental capacities must be equally and similarly different from normal individuals from a neurological point of view. Second, psychiatrists have to cure mental disorders by acting on the nervous systems of mentally ill people so that mentally ill people may turn out to have mental capacities falling within ranges of variation that can be considered normal for humans. The universal treatment thesis, it seems to us, follows from the conjunction of these two assumptions with the received view.

In this chapter, we argue that the received view may be accepted and that the first implication just mentioned follows from it, but we also claim that the second implication does not follow, because the first complementary assumption can be rejected on evolutionary grounds. In the second section, we briefly discuss the received view and its relevance for the links between psychopathology and ethics; we suggest that the notion of "function" plays a central role in this respect. In the third section, we try to show that the universal treatment thesis is a widespread view, but we also suggest that it is unacceptable, because it purports that the notion of a function could be defined in relation to the whole species only. From an evolutionary perspective, though, it seems that the notion of a function must be defined in relation to both an individual and the species to which it belongs. In the fourth section, we consider some data taken from pharmacogenetics and from psychopharmacology, and we try to explain why psychiatric treatments have to be molded for individuals, not for the entire species, contrary to the universal treatment thesis. We conclude with some remarks concerning the ethical bearings of our claims.

It is worth noting that this chapter is intended as a contribution to the evolutionary approach to pathology. Although a contender in medical debates since Darwin's times (cf. Corbellini 1998), the evolutionary approach to pathology became an autonomous field of study and research during the early 1990s, especially through the work of the psychiatrist Randolph Nesse and the biologist George Williams (Nesse and Williams 1991, 1995). Several collections of essays have subsequently been published on the topic (cf. Donghi 1998; Stearns 1999; Trevathan, Smith, and McKenna 1999; Canali and Corbellini 2004), and mental disorders have also been a focus of this perspective (cf. Stevens and Price 1996; McGuire and Troisi 1998; Canali 2001).

The main contention of the evolutionary approach in medicine, which we want to support, is that each individual is, at least partially, the expression of a particular genetic program and that this program is a historical and unique product of evolution, which was molded by the mechanisms of phylogenesis (i.e., genetic variation and natural selection). Therefore, according to evolutionary medicine, epidemiological phenomena, specific individual vulnerability to particular diseases, the ways and timing in which each individual reacts to a pathogen, falls ill, or regains her health, all depend *also* on historical, phylogenetic processes. In this view, an exhaustive explanation of pathology cannot be limited to the immediate causes triggering a pathological process, but it must also take into account the remote causes, that is, it must make use of evolutionary explanatory categories.

More recently, the evolutionary approach was also applied to the analysis of drug consumption, with the hope of helping the prediction of therapeutic effectiveness – particularly in the case of antibiotics, antivirals, and the treatment of cancer through chemotherapy (cf. Davies 1996; Levin and Anderson 1999; Normak and Normak 2002). Surprisingly, very little work has so far been done on psychopharmacology, despite the fact that the practice of psychiatry seems to look at the Darwinian paradigm with much more sympathy than any other medical specialty does. Therefore, this chapter should be considered as a contribution both to philosophical reflection and to evolutionary psychopharmacology.

PSYCHIATRY, ETHICS, AND FUNCTIONS

Let us now give a closer look at the received view and at its relations to ethics. As we have seen, the received view claims that, in psychiatry, an individual (or a state) is pathological if and only if it fails to be as an individual (or a state) of that sort *ought to be*. This means that sortal concepts allow us to group together individuals (or states) in virtue of some of their characteristics; those characteristics must belong to all the thus grouped individuals. A pathological individual (or state) will then be one that falls under a certain sortal concept but fails to have one or more of the properties that individuals (or states) falling under that concept must have. It might be objected that sortal concepts must be clear-cut and that it makes no sense to claim that an individual falls under one of them, while failing to have one or more of the properties that individuals falling under that concept must have. Let us think about the case of humans and mental capacities. Either an individual has all the relevant mental capacities and thus falls under the concept human, or that individual lacks one of those capacities and cannot then be called a human.

It can be replied, though, that sortal concepts need not be so clear-cut. The reasons why we claim that different individuals fall under a certain concept may vary, and our criteria for grouping individuals may differ from individual to individual even within the same group. We can claim that both *a* and *b* are humans, but on different grounds. For example, we may talk to and discuss something with *a*, and this is a good enough reason to call him a human, but we might not be able to do the same with *b*, who is in a coma. However, we can still claim that *b* is a human, if we know his parents, and we know that they are both humans. Conversely, we might not be able to apply this latter criterion to *a*, if we do not know her parents, and, had we not possessed other good grounds to trust her humanity, we could have reasonably wondered whether they were aliens. Thus, it seems that different things can be claimed to fall under the same sortal concept on different grounds, but, still, sortal concepts tell us what properties should ideally have things falling under them (cf. Strawson 1959; Wiggins 2001). Sortal concepts, then, are semantically linked to the stereotype, which we associate with the things falling under them.

It is worth stressing that the identification of the properties linked to a certain concept and claims of sortal identity about individuals do not constitute a circle. It could be objected, indeed, that one needs to know what the necessary properties required by a sortal are, before claiming what individuals fall under it; yet, at the same time, one needs to know what individuals fall under a certain sortal concept, before understanding what the necessary properties linked to that sortal are. The reply offered in the previous paragraph can be readjusted for this objection: sortal attribution is not a clear-cut process, but it is a process that involves grouping together individuals on different grounds, and subsequently realizing that they can be grouped together because they all conform to some standards, at different degrees. When we start reflecting on these standards, we come up with the idealized set of properties, which an ideal individual falling under that sortal concept should have. This is not to say, however, that the idealized set of properties cannot be changed: the more individuals falling under that concept we encounter, the clearer can we be about what properties are necessary for individuals of that sort.

This account does not attempt to say *how we should think* about the world, but it describes *how we do think*. In Strawson's terms, it is descriptive, rather than revisionist. It is in virtue of the fact that we do think "sortally" that we can make sense of the idea that something is pathological, that is, that it is not as it should be.

We can now turn to the second point that needs to be discussed in this section: the reason why the received view entails the fact that psychiatry is related to ethics. The point is that, as noted, the concept of pathology is normative, and normative concepts may be the ground for ethical norms. Why should this be so? One way of explaining the link between normative concepts and ethics could be the following. In the case of living organisms, the standards that set what an ideal member of a species is like also set what the correct functioning of such an individual is. This is a clear consequence of the normative character of sortal concepts: different living beings may be grouped together under a certain sortal concept, that is, they may be considered a species, because they can live the same sort of life, that is, they have functional parts that enable them to perform specific operations. Naturally, the species sets the standards for the correct functioning of its individuals, but different individuals may conform to those standards at different degrees. However, performing their function at their best is what they all ought to do.

At this point the reader may worry that this approach is naively open to an objection that is commonly advanced against attempts to explain normativity through the notion of function. It may seem, indeed, that the view that we are suggesting rests on an equivocation between two senses of "function"; the unquestionable claim that living organisms have functionally organized parts does not entail, let alone is it identical to, the claim that each of those organisms as a whole has a function. Parts are functional because they contribute to the life of the organisms to which they belong, the objection goes, and thus their contribution may be what they ought to do; but entire organisms cannot be said to have functions because they are not parts of bigger wholes. Recently, however, several philosophers have persuasively attempted to respond to this objection and to contend that entire organisms have functions. It is not necessary to get into the details of these debates here, but it is worth mentioning at least a few attempts. Murphy (2001), for example, has argued in favor of the principle of function compositionality, according to which the parts of a whole may have a function only under the condition that the whole has also a function. Murphy's point is that we cannot make sense of the idea that a part has a function unless we consider it as contributing to an activity of the whole to which it belongs. That activity, though, is the correct functioning of that individual and, thus, is its function. Foot (2001, pp. 25–37) has also suggested that the natural function of an organism is the organized activity that all the organisms of that species perform when they live: "All the truths about what this or that character does, what his purpose or point is, and in suitable cases its function, must be related to [its] life cycle. The way an individual *should be* is determined by what is needed for development, self-maintenance, and reproduction: in most species involving defense, and in some the rearing of the young" (Foot 2001, pp. 32–33).

One could worry that the appeal to functions makes attempts to explain normativity of this sort intrinsically incompatible with an evolutionary approach to these matters. The reference to functions, in fact, seems to suggest that things having functions have purposes, for example, they are made for some end or other. But, according to neo-Darwinism, there is no purpose in nature: the existing forms of life are not designed to be how they are, and their functioning is the mere result of casual systemic variations and fitness to the environment (cf. Ayala 1972). It must be recognized, though, that the notion of function referred to by the previously mentioned theories does not require any purpose, or any design. In order to explain how this may be so, let us consider Bedau's (1992) distinction among three grades of teleology.

Grade 1 is the sort of teleology in which an end benefiting an individual (e.g., favoring its survival or reproductive chances) is arrived at by chance, in the sense that it was not the intended aim of an agent or the result of a normal etiological chain characterizing the functioning of that individual. In grade 2, though, the end of a teleological process is good for the individual and is arrived at through an etiological process characterizing the functioning of that individual, although its goodness did not play any causal role in that process. In grade 3 teleology, the end is good, is arrived at through a proper etiological chain, and the goodness involved does play a causal role in the process. This is the case of intentional action: the content of a representation in the agent's mind, the realization of which is the end of the process, is part of the causal conditions that give rise to the process.

The notion of function involved in the previously mentioned theories of normativity requires only grade 2 teleology. If a living thing has a function, there is something that is good for it, namely, the correct performance of its proper activity. Thus, the condition for grade 1 teleology is satisfied. However, the preceding views also claim that living organisms have functions because they are included in a certain species – that is, in virtue of the fact that they have a structure that makes them fall under certain sortal concepts. The performance of the function proper of their species, thus, etiologically depends on their organic structure, and the further condition for grade 2 teleology is also satisfied. However, the normative views mentioned previously do not claim that the organic structure of the individual that belongs to a certain species is the result of intended action, or that it is an end chosen by a designer. In fact, they do not need to affirm that the function has any causal role in determining the organic structure typical of the individuals belonging to that species. That structure may well be the result of evolution. As a consequence, those normative views do not need grade 3 teleology and are compatible with evolutionism. Those views, indeed, attempt to ground normativity on nature.

A last remark about the notion of function involved here. A living thing *i* falling under the sortal concept *C* has the function *F* in virtue of the fact that things falling under *C* ought to have a certain set of properties *P* (which includes both mental and physical properties and constitutes the stereotype that we associate with things falling under C). The individuals having the properties contained in P, in fact, in virtue of those properties, are suitably structured to live a certain kind of life, which is their function F. Thus, the function F of i is a function of the properties included in P, which make i a C.

So far, we have seen that living things have functions, and that what functions they ought to have depends on the sort of things they happen to be (i.e., on some properties they must have in virtue of the fact that they fall under a certain sortal concept). This means that the notion of a function is a value notion, that is, it is good for an individual to function well. This value notion can probably explain the human moral capacity and – possibly – ethical systems: the human moral capacity is the set of cognitive and emotional abilities that allow a well-functioning human being to live well and to fit in a society; an ethical system is a set of norms governing a human society that maximizes the chances of flourishing for each involved individual. However, can the notion of a function help justify ethics? Can it tell us why we should do what is good for us? Or for some other living thing?

We cannot deal with this issue here, and our focus will be mainly on the problem of the explanation of ethics. However, it may be worth noting that the previously mentioned theories give different but probably entrenched and compatible, justifications of morality by appealing to the existence of the functions of living beings. Murphy claims that the attainment of her own function is what a human being ought to do, because the same state of affairs, which is her well-functioning, can be both the content of her theoretical reason (which grasps it as her well-being) and the content of her practical reasons (which considers it as a reason for action). Foot, instead, takes over Davidson's famous distinction between two kinds of reasons for action: those which are reasons relative to a certain consideration, and those which are reasons all things considered. Only the latter are moral reasons. And considerations on what a human should do, on what it is to behave well for her, gives us precisely reasons *all things considered*.

We can now finally turn to psychiatry. Within a conception of normativity that grounds norms on nature in one of the manners considered here, psychiatry is relevant for ethics because it can help to understand what the correct functioning of a human being is. One of the things that psychiatrists do is to define normal mental behavior and to distinguish it from pathological mental behavior. The correct functioning of a human being, which constitutes a reason for an action and grounds ethical behavior, includes also different sorts of mental activities. Thus, psychiatry can contribute to determine how a human being ought to be.

Of course, psychiatry is not the only or the preferred foundation for ethics. As we have seen, sortal concepts have fuzzy borders and their application may involve the deployment of several criteria. To decide what a human ought to do in a moral sense – what his *all things considered* reasons for action are – one needs to take into account all these criteria, and to balance together different opposing reasons to claim that a living thing falls under a certain sortal or that things falling under a certain sortal must have certain characteristics. Psychiatry delivers some of these reasons, which need to be weighed against those coming from other fields of experience. Psychiatry is an autonomous discipline, which rests on its own grounds and does not need to consider other points of view. But its results can be used to try to determine how a human being should be – *all things considered* – and it is at this point that its results have to be weighed against the conclusions of other fields. At that stage, though, we have already abandoned psychiatry to enter the domain of ethics.

There is at least another way in which psychiatry and ethics interact. As we have seen, psychiatrists try to understand what is a normal human mental behavior, and, in order to do that, they must try to figure out what mental properties and capacities an individual must have in order to fall under the concept *"human."* Given the previously mentioned autonomy of psychiatry as a scientific field, they will try to do that on purely empirical grounds, in particular in virtue of evidences coming from neurobiology, genetics, cognitive psychology, and the like. Once a provisional understating of what normal human mental behavior is (namely, what set $M = <m_1, m_2, \ldots, m_n>$ of mental properties an individual i must have in order to fall under the sortal concept H, which is the concept human), abnormal cases can be identified, and means of treatment looked for. The rationale beyond the behavior of psychiatrists includes at least the five following preconditions: psychiatrists realize that there is a sortal concept H, which allows us to group humans together; they believe that a standard individual falling under H must have the set of mental properties $<m_1, m_2, \ldots, m_n>$; they encounter an individual that they take to fall under the concept human, although he fails to have at least some of the properties $<m_1, m_2, \ldots, m_n>$; they take it that, as an H, i should have all the mental properties $<m_1, m_2, \ldots, m_n>$, and thus they qualify i as an abnormal, ill, or deviant H; and they look for a "cure" for i, that is, a means to render i as "normal" as possible, and attempt to grant him as many as possible of the properties included in the set $<m_1, m_2, \ldots, m_n>$.

The following consideration may explain why this must be the rationale beyond psychiatric practice. We all normally believe that a psychiatrist trying to heal a dog for failing to read would be insane, whereas a psychiatrist trying to heal a dyslexic human for the same reason would be absolutely rational.

This difference among our judgments can be best explained if we accept the truth of the foregoing five preconditions. Both the dog and the human fail to have a certain mental capacity, but we think that the human should have it, whereas the dog is just not the kind of being that should be expected to have it. Thus, it makes no sense to try to make a dog read, whereas it seems to us absolutely mandatory to try to help the human.

The link between psychiatry and ethics lies in the fact that the set M is a subset of the set P, which includes all the properties (mental and nonmental) that an H should have. For the autonomy of psychiatry, M should be determined by psychiatry, on its own grounds. But when it comes to deciding whether and how to cure an individual i, that is, to enforce on i as many as possible of the properties contained in M, a problem arises. The psychiatrist must consider whether each of the properties in M (or each of the consequences of a possible cure aiming at a maximization of the properties in M) can coexist with the other properties in P, when considered overall. If there are inconsistencies among some of the properties, the conflicting properties must be weighed against each other, and the prevailing one will deliver a reason for action. This means that psychiatry can give reasons for action that are relative to psychiatric considerations, but these have to be weighed against reasons coming from other sources, in order to find out reasons that hold, *all things considered*. This implies that when deciding how to treat a certain individual, psychiatric considerations have to be balanced with considerations having to do with what the function of a human being is. With this step, the psychiatrist abandons her own field, and enters the realm of ethics.

As a consequence of these remarks, we can claim that notions like "normality" and "pathology" are not dispensable. They lie as a rationale beyond the sorts of (the psychiatrist's) behavior, which we consider normal and rational. However, they need to be flexible, because sortal concepts, which are their preconditions, have fuzzy and unclear borders. Our evidences for determining the set P and its subset M may vary, and this might eventually push us to revise our conceptions of normality and pathology. This can be called "epistemic flexibility," because it depends on our epistemic standpoint. There may also be, however, a kind of ontological flexibility: evolutionism teaches us that species may change across history, and thus we may need to revise our conceptions of normality and pathology due to changes in the very structural and organic organization of the standard individuals of a species. In the former case, the flexibility depends on our limits in determining what properties an individual ought to have in order to fall under a certain sortal concept. In the latter case, there are real changes of the properties that are necessary in order for an individual to fall under a certain sortal.

The upshot is that the received view of psychopathology entails this requirement: an adequate psychiatric practice needs to consider ethical questions while deciding what mental properties are required for a normally functioning human being. In order to do this, psychiatry has to be both epistemically and ontologically flexible. Epistemic flexibility requires that psychiatrists recursively remold their notion of mental normality, through wider considerations about what the function of human beings is (i.e., what the required properties of a standard *H* are), by considering fields of experience other than psychiatry, and by observing new cases of *H*s. In this case, the set *M* can be changed on the grounds of reasons coming from other subsets of *P*, or on the grounds of previously unconsidered traits of *H*s, which can be highlighted by a newly encountered *H*. Ontological flexibility requires that psychiatrists recursively remold their notion of mental normality by considering whether cases of nonstandard individuals should be taken as deviant cases or as the marks of a shift in the history of the species.

The universal treatment thesis (i.e., the idea that each mental disorder must have one perfectly appropriate cure, which scientists have to work out and clinical psychiatrists have to apply to individual patients) does not seem to be either epistemically or ontologically flexible. It is not epistemically flexible, because it rests on the false idea that sortal concepts can at least in principle be clear-cut and, thus, that a universal characterization of normality and pathology can be defined. It is not ontologically flexible because it overlooks the contribution of evolutionism and does not consider the possibility that species may change; indeed, it does not consider that treatments may be made inadequate by the evolution of the species, even if they were perfectly efficient when they were first worked out.

It is our opinion that an evolutionary approach may render psychiatry suitable to meet the ethical requirements set by the received view.

THE UNIVERSAL TREATMENT THESIS AND ITS PROBLEMS: THE FUNCTIONS OF INDIVIDUALS

The Evolution of Psychotropic Drug Consumption

The wide acceptance of the universal treatment thesis is witnessed by data concerning the drug consumption relative to the treatment of certain mental diseases. These data reveal what the ordinary attitudes toward those diseases are. After considering this example, we discuss how an evolutionary approach could lead to very different attitudes toward those diseases.

Table 6.1. *The Psychotropic Drug Market in Italy (million sold)*

	1995	1996	1997	1998	1999	2000	2001
Antipsychotics	11,675	12,432	12,468	12,940	13,428	13,114	13,267
Antidepressants	16,750	18,275	19,413	20,524	22,493	24,435	28,380
Tranquillizers	61,107	64,051	63,464	63,292	63,497	63,267	63,608

Source: World Health Organization.

Recent evidence indicates that the way psychotropic drugs are prescribed in the United States (Pinkus et al. 1998) and in Italy (Pani 2000) is changing. In the United States, in the ten-year period between 1985 and 1994, the number of psychotropic drugs prescribed rose from 33 to 46 million. In the past few years, tranquillizers or hypnotics, which had previously been the most frequently prescribed drugs, have been overtaken by antidepressants, which have doubled in quantity, reaching over 20 million prescriptions in the past five years; the use of stimulants and "tonics" increased by 500 percent over the same period. Similar trends were observed in Italy (Table 6.1), where prescriptions were grouped into three broad categories: tranquillizers, antidepressants, and antipsychotics. The data show a large increase in antidepressant prescriptions, a smaller increase in antipsychotic prescriptions, and an even smaller one in tranquillizer prescriptions.

These data seem to suggest a possible generalization concerning industrialized countries. These countries are characterized by several areas in which a rapidly changed environment presents traits of evolutionary "novelty" that may be deemed significant. The effects of these new conditions on individuals seem to be the emergence of psychic unfitness, as the constantly rising numbers of people requesting psychiatric treatment seems to testify. The efficiency of these treatments, though, seems quite dubious, because the situation of mental diseases in industrialized countries seems to be far worse than in developing countries (Brown et al. 1998).

These treatments, furthermore, seem to follow the "universal treatment thesis," since they suggest that cases of psychic malfunctioning are increasingly coped with using standard pharmacological remedies. Because these remedies prove to be inadequate, though, the universal treatment thesis should be reconsidered. Our contention is that the problem with the universal treatment view is that it does not take into account the evolutionary meaning of some mental diseases. Were the evolutionary meaning considered, the universal treatment thesis would be abandoned, and the resulting view would be more flexible in the senses required by the ethical demand we have considered, and, probably, more effective on the psychiatric consequence of the evolutionary mishmash created by contemporary industrialized societies.

Some examples show that the universal treatment view is insensitive to the demands of evolutionism. Before considering them, though, we should consider some argumentative patterns that evolutionary theory could provide to medical and, in particular, psychiatric thinking.

Evolutionary Explanations in Psychiatry

According to a recent analysis of explanation in pathology (cf. Nesse and Williams 1991, 1995, 1999), evolutionary models to account for psychiatric disorders may be grouped into seven categories.

ADAPTATION AND DEFENSE. Several pathologies or organic weaknesses actually act as sensitive defenses and adaptation mechanisms. In psychiatry, they amount to the reactions of alarm and fear. These apparently dysfunctional conditions may in fact be adaptive and aimed at preparing the individual to cope with stimuli relevant for his survival and for that of the species. Depression may be considered a defense mechanism of this kind, aimed at inducing the implementation of a detachment or a break with the past, and the reconstruction of novel forms of adaptation (cf. McGuire et al. 1997; Nesse 2000).

CONFLICT WITH OTHER EVOLVING ELEMENTS. One specific case is that of the conflict between parents and children, ranging from the pathogenic potential of the fetus on body and neurochemical equilibria of the mother, to the load of elements with possible pathogenic valence, which are implicit in the caring for children and in relations with them. The social environment can also engender a series of potentially pathogenic conflicts, because it induces both competition among individuals and the emergence of various relational systems.

EVOLUTIONARY MISMATCH. Our bodies and our behavioral reaction patterns, which evolved slowly during our ancestral life in the savannah, are no longer adapted to the environmental and social contexts of the modern age (Eaton et al. 1988). One striking case is that of substance abuse. The epidemic of drug addiction related to the modern age can be interpreted as the result of exposing human beings to pure psychoactive principles toward which the human nervous system is currently unable to provide adaptive responses (Nesse and Berridge 1997). Moreover, although the emotions are adaptive tools that can be used to cope with situations relevant for survival, it is also possible that they are elicited by erroneous evaluations of stimuli and that their expression threshold, defined by means of slow selective processes, can no longer cope with the transformations produced by man in his environment (the enormous number of cognitive stimuli and the rules governing social behavior). This is true, for example, of the fight reaction, which in contemporary society is produced by a large number of stressful or frustrating situations, but cannot be expressed at the behavioral level, for sociocultural reasons (Marks and Nesse 1994; Nesse and Young 2000). Eating disorders can also be accounted for with patterns whose explanation can be included in this category (Neel et al. 1998).

EVOLUTIONARY TRADE-OFFS AT THE GENETIC LEVEL. A number of pathological conditions are the results of specific genetic adaptations to a given environment. A gene may afford several advantages in specific environmental contexts, but, at the same time, it may increase the likelihood of developing certain diseases. The most striking example in psychiatry seems to be the bipolar or manic-depressive disorder. This disorder seems to have a highly hereditary component, actually believed to be as high as 50 percent. The conservation of a gene that causes a disorder that can sometimes be highly incapacitating may be explained in virtue of the fact that its presence may bring about advantages, which offset any negative effects. Several studies suggest that people suffering from manic-depressive disorders are more creative, more enterprising, and better in achieving social success than ordinary people; that is, they are bearers of behavioral traits that can ensure a reproductive advantage. Consequently, the gene or gene combination responsible for this psychiatric disorder thus maintains a high frequency (Goodwin and Jamison 1990).

EVOLUTIONARY TRADE-OFFS AT THE LEVEL OF THE COMPLEX PHENOTYPIC TRAITS. Each somatic or behavioral trait in an individual is the expression of a complex genetic and epigenetic equilibrium between somatic structures and psychological functions. Some genes are simultaneously part of the biochemical systems governing different processes – for example, organ development, hormone synthesis, or a specific enzymatic reaction. In this way, a genetic mutation that increases the efficiency of one function may jeopardize the effectiveness of another biological activity of a behavioral program and, thus, expose the organism to the onset of specific diseases. For example, the strong reactivity of the cardiovascular system to emotional stimuli may increase vulnerability to disorders in these organs. On the other hand, were the cardiovascular apparatus less sensitive to the stimuli that can trigger affective reactions important for the survival of the individual and the species, the organism could turn out to be inadequate to cope with risk situations or even to reproduce.

HISTORICAL CONSTRAINTS AND DEPENDENCE ON EVOLUTIONARY PATH-WAYS. Evolution proceeds by recycling and coadapting the "old" biological and psychological material of the species. The best trade-off between materials and preexisting biological functions can hardly coincide with the best and most effective solutions that could theoretically be elaborated for a functional structure. An example could be the conditioning determined by the affective responses on cognitive faculties. This interference can sometimes be quite pathogenic. This mishmash depends on the fact that the development of cognitive faculties was superimposed on a consolidated inheritance of emotive patterns.

RANDOM FACTORS. The evolutionary process does not follow any prearranged pattern of development aiming at maximum efficiency. It uses, adapts, and remodels existing functional apparatuses and anatomical parts (a form of biological tinkering) and is largely the result of the action of random factors (genetic variations, environmental modifications, changes in ethological relationships, and so on). Evolutionary randomness alone would be sufficient to account for many human diseases. In this sense, "randomness of drug prescriptions" adds a further contextual variable of considerable interest. In the case of psychotropic drugs, randomness may depend on the therapeutic options of different therapists or on a single therapist changing his mind in a short period of time. Frequent and rapid changes from one molecule to another may be the result, which leads to predictable but unpleasant consequences, such as withdrawal symptoms, and sudden change and necessary adaptation in receptor interactions, signal transduction, and even in gene transcription. Random variations entails the reading of nonrandom genetic programs, which were selected for other purposes and had evolved in response to different stimuli. A nonunitary (dimension- or category-based) nosography amplifies the impact of nonunivocal therapeutic decisions, that is, decisions that are not dictated by homogeneous working hypotheses and models of the health-disease continuum. Indeed, there are several therapeutic approaches available (dynamic, behavioral, familial, or biological) that have resisted numerous attempts of unification or integration.

With these evolutionary explanatory patterns in mind, we can now consider some examples, in order to emphasize the faults of the universal treatment thesis and the advantages of an evolutionary approach.

Psychotropic Drugs and the Adaptive Significance of Psychiatric Symptoms

The evolutionary approach suggests that drug therapy should take into account the adaptive significance of certain psychiatric symptoms (McGuire and Troisi 1998). Historical constraints and the dependence on evolutionary trajectories have actually led to the cognitive processing of external stimuli and the assessment of the personal condition vis-à-vis the surrounding context to be largely based on emotional processes, which are heavily influenced by the affective dimension (Damasio 1994; LeDoux 1996). From this perspective, the use of tranquillizers and antidepressants to treat subclinical conditions, for subthreshold action, or to act on the penumbra of mood disorders may jeopardize the adaptive function of certain emotional responses. Let us consider two cases: first, the relation between the emotive information and the motivational drive; second, medicalization of character and generalization of temperament.

EMOTIONAL INFORMATION AND MOTIVATIONAL DRIVE. A certain degree of anxiety is quite physiological insofar as it can signal a danger or threat, which has not yet been processed and perceived at the conscious level; alternatively, it may reinforce the motivation to act functionally once the state of awareness has been reached (Nesse 1999).

Several depressive symptoms have a similar adaptive function, both informative and motivational. They can signal an existing gap between expectations or investments and results, or a profound clash between one's personality and the condition in which one lives. They can lead the subject to break off the investment and to abandon the situation, halting the process in order to recover and to work out a fresh strategy.

Subthreshold intervention may consequently interrupt the information flow from the deepest levels of the brain to the cortical areas and thus hinder or prevent the cognitive processing of the problem, or the search for a suitable solution. Likewise, an incorrect use of drugs may inhibit functional responses and motivational drives aimed at eliciting more appropriate behavior. Paradoxically, drugs may contribute to maintain the pathogenic situation.

This problem has important psychological and social bearings. Therapeutic abuse can in fact force individuals to adapt to existential situations and contexts, which are objectively painful or in any case display a discrepancy in respect to character profiles. This acceptance or acquiescence can have severe repercussions on individuals and on society. Chronic pathogenic factors will continue to affect the former, while other unacceptable conditions will probably continue to be maintained and reproduced in society.

Certain drugs used to treat mood disorders change the expression of emotions by the patient and thus may suppress an important way of ensuring interaction with and modification of the external environment. Consequently, this could affect the behavior of the individuals who surround the patient and possibly modify their understanding and willingness to help (Lewis 1934). In this sense, the treatment could modify and sap the effectiveness of some of the relational and social factors, which would normally help the recovery and the reconstruction of meaningful affective and social relations.

MEDICALIZATION OF CHARACTER AND GENERALIZATION OF TEMPERAMENT. Similar remarks may be spelled out about the increasingly widespread use of antidepressants to modulate mood, and to correct "character flaws." This trend seems a real attempt to medicalize character, in order to conform it to socially prized models. The social phenomenon is analogous to the boom of aesthetic surgery to conform to socially accepted standards of physical beauty (Kramer 1997; Knutson et al. 1998).

This trend may increase interventions in cases of subthreshold symptoms, in the face of subclinical disorders, and this may eventually lead to the generalization of temperament and the leveling down of individual differences. This scenario would have various hidden risks. Temperament, as the genetic expression of personality, is strongly predetermined and its alteration by means of drugs may cause a strong conflict between genetic inheritance and a chemically modified phenotype. (Of course, we are referring to cases in which there is no ongoing disease.) The following problem then arises, because certain character traits are hereditary.

Let us imagine an individual who transmits to his descendants a genetic endowment with a proneness to fear (a phobic vulnerability). Let us also imagine that the same individual, under antipanic drug treatment, displays a nonphobic phenotype to his children. Even if this may be desirable in a sense, we cannot predict the effects of this apparently irremediable clash between a (genotypic) Darwinian inheritance and a (phenotypic) Lamarckian one. Will the children too be obliged to be medicalized or will the example be sufficient to prevent the phobic potential of their genome from being expressed? And if this were the case, would this not paradoxically show that nonbiological variables are able to influence the expression of the genetic endowment and that, therefore, behavioral desensitization therapies, in such cases, would be preferable to the use of psychotropic drugs? Around this issue, there is currently a very heated debate.

In these two examples, the application of the previously considered evolutionary explanatory patterns (which are here entrenched in ways that we leave to the reader to work out) leads to the conclusion that medicalization ought to be avoided. Were the evolutionary considerations overlooked, the output would have been the opposite. The point is that the evolutionary approach is more flexible, in both the senses considered previously, than the alternatives. The considered patterns of explanation, indeed, allow us to consider the evolutionary significance of certain mental traits: they show us some of the possible processes that make us remold the set M of mental properties that we stereotypically associate with humans, through the consideration of other intuitions of ours about what humans have to be from a nonneurological standpoint (i.e., through the consideration of all the properties contained in the set P). As we have seen, the set M cannot be medically molded purely on the grounds of psychiatric evidences but must be shaped according to considerations concerning other reason-giving characteristics of humans, which cannot be accounted for by psychiatry itself. The evolutionary explanatory patterns considered here offer a substantial amount of these considerations.

These patterns of explanation, as we noted, are flexible in both senses, epistemic and ontological. They may bring in considerations relevant for the remolding of M, and these considerations may depend either on the focus on previously unconsidered data or on the realization that evolution caused a shift in a species. However, one could claim that all this is compatible with the universal treatment thesis: cannot the evolutionary considerations be conjoined with the universal treatment approach? The answer is the negative, and this is why, while considering the notion of flexibility, we noted that the remolding of M requires both considerations concerning other nonpsychical stereotypical characteristics of humans, and questions concerning the stereotype associated with the concept H, which may originate from the encounter with deviant Hs. It seems to us that evolutionary thought offers a characterization of the relations between an individual and the species to which it belongs that offers the desired sort of flexibility, but which is incompatible with the universal treatment thesis. As we noted in our introduction, evolutionism makes the individual central, more than alternative approaches do. In the next section we will see why.

WHY FUNCTIONS ARE ALSO MATTERS OF INDIVIDUALS

Evolutionism, Pharmacogenetics, and Pharmacogenomics

In order to explain the role of individuals in sortal identification, we can focus on the case of response to drugs. Genetic variability, the raw material of phylogenesis, is obviously expressed also in the specificity with which each individual responds to drugs. One aspect of genetic variability at the population level is represented by genetic polymorphism. As a source of variability, polymorphism is filtered through natural selection and is thus functional to adaptation to environmental and ethological changes; that is, it is neutral but in any case essential to phylogenetic transformation.

At the pharmacogenetic level, polymorphism is important for genetic flexibility, which has made it possible for organisms to cope in the encounter with new substances and probably represents one of the most complex expressions of the coevolution of the animal and plant kingdoms. A large number of genetic polymorphisms of pyschopharmacological interest are known today; of course, they are relevant for clinical psychiatry (Kalow 2001). One of the better known is that linked to the polymorphism of an element in the Cytochrome P450 hepatic enzymatic system, CYP2D6, identified for the first time as responsible for the variation in the metabolism of debrisoquine (an antihypertensive). Dozens of genetic mutations are now known to be associated with this polymorphism. CYP2D6 is one of the most important enzymes involved in the oxidative metabolism of drugs and catalyzes the oxidation of several dozen drugs, about 20 percent of all commonly prescribed substances. The list of CYP2D6 substrates is a long one and includes all the tricyclic antidepressants; several serotonin reuptake inhibitors, such as fluoxetine and paroxetine; as well as many antipsychotics, such as haloperidol, perphenazine, and risperidone (Kalow 1991; Bertilsson 1995; Ingelman-Sundberg et al. 1999). In individuals with weak metabolization all these drugs reach concentrations from two to five times greater than in normal metabolizers, which implies that in more serious phenotypes the recommended dosages can lead to toxic concentrations.

Very relevant for clinical psychiatry and psychopharmacology is also the polymorphism of another element in the hepatic Cytochrome P450, CYP2C19, which affects about 3 percent of white Caucasian individuals. CYP2C19 metabolizes several drugs in psychiatric use such as imipramine, diazepam, citalopram, and amitryptiline, which are consequently affected by this polymorphism.

Other important pharmacogenetic polymorphisms in psychopharmacology are those of the receptors with which the psychotropic drugs interact (Masellis et al. 2000).

These facts indicate that pharmacogenetics provides molecular evidence to corroborate the idea that an individualized approach to therapy is required (Brockmoller 1999; Ozdemir et al. 2002). The evolutionary conception of medicine calls for such an individualized approach, because evolutionism takes each organism to have an irreducibly individual nature, in virtue of the idea of population on which Darwin founded his doctrine.

Specificity of Drug Action and Integrative Aspects of Biological and Adaptive Functions

Research on psychotropic drugs and their therapeutic use is increasingly being concentrated on substances characterized by a highly specific action. These are pharmacological principles capable of acting selectively on individual neuronal systems or, even better, capable of modulating the functioning of specific receptors in the same neuronal system.

This approach, although innovative, might not be particularly effective, unless the molecule selected displays an excellent affinity and an extreme specificity for a single target of known physiology and with a proven role in the pathological process that is to be treated. One striking example of the difficulties encountered in this respect by modern psychopharmacology is given by current schizophrenia therapies. Molecules that may have radically different action mechanisms are today available to clinical psychiatrists. Several of these (e.g., Clozapine) have a multireceptor profile – that is, they can act on several different molecular targets. Others (e.g., Amisulpride) act on a single receptor. For example, in the past fifty years, dopamine D2 was shown to play a role in the physiopathology of schizophrenia. However, a unifying account of how both these drugs act has still to be found.

More recent research has moved toward the identification and development of substances with a capacity for action at gene level or at that of genetic networks, which specify and modulate the functions of various functional apparatuses of the nervous system.

Nevertheless, the final therapeutic effect on a patient depends on numerous additional factors: pharmacogenetic individuality – mentioned previously – and peculiar traits of individual constitution. The latter can be related to complex genetic and metabolic factors, which may not depend only on parts of the genome directly responsible for the functions of the nervous system but can still affect the pharmacodynamics and pharmacokinetics of drugs (Hofbauer and Huppertz 2002). The important point is that these genetic and metabolic factors are products of evolution, that is, the result of a mutual coadaptation, which may depend on various phylogenetic compromises and may be influenced by evolutionary trajectories.

The evolutionary approach also highlights the integrated nature and the mutual adjustment of the functional apparatuses of an organism. In this sense, one important but usually underestimated factor in the determination of the long-term action profile of a drug is the reaction of the regulatory systems of the organism to the action of the substance itself. This is a compensatory response, which tends to restore the state of equilibrium, either functional or pathological (i.e., the homeostasis preexisting the drug's action), and thus generally amounts to the reduction or suppression of possible therapeutic effects. Occasionally, it may even induce adverse effects. From a theoretical point of view, it is impossible to predict the precise "readaptation" reactions to a drug on the basis of the profile of its receptor. It is quite reasonable to expect that the magnitude of the side effects will be proportional to the number of the action sites of the drug. But this does not tell much about its therapeutic effects. The intrinsic limit to pharmacogenomics lies in the possible confusion of psychotropic drug "safety" acquisitions with those of their "efficacy," two dimensions of their action on the organism that are very different from each other.

The Individual Function

These examples can be used to make a general point and also a point specific to psychiatric therapy.

The general point is that the stereotypical set of properties that we associate with a certain concept involves largely phenotypic properties. A certain individual, though, belongs to a certain species in virtue of its genotypic inheritance. Thus, variation in the environment or polymorphisms may cause, concerning a certain individual, a divergence between its belonging to a species and its falling under a certain sortal, the former being ascribable on genotypic ground, the latter on the fitting of the phenotype in the stereotype associated with that sortal. Given the relationship between the normal function of the individuals falling under a sortal and the set of stereotypical properties associated with that sortal, the possibility of this divergence entails that a certain individual may be inapt to function as members of its species are stereotypically believed to function. A polymorphism or a change in environmental variables may cause the expression of a phenotype that cannot function as individuals of its species are normally taken to have to function, notwithstanding

116

the fact that it certainly belongs to a given species. This leads to the need of a new notion, that of an *individual function*, namely the proper functioning of an individual. This is a function of the stereotypical function associated with its species and of its individual phenotypic traits, depending on environmental and polymorphic variables. For example, in order to know what the normal functioning for a certain human being may be, we need to consider what the stereotypical functioning of humans is, and how that particular human being differs from a stereotypical phenotype.

The notion of the individual function involved in the evolutionist approach shows that this approach is epistemically and ontologically flexible and that it explains how the consideration for individuals can help individuating the set M of mental properties that an individual *should* have.

The more specific point is that the need for the notion of the individual function shows that the universal treatment thesis is inadequate. Psychiatric treatment must consider a number of individual features, through the considerations of the evolutionary significance of several traits of the individual, even apparently pathological ones, before any therapeutic measure can be decided. There is no room for generalizations and for the application of standard models of treatment to all individuals, because even what counts as pathological may vary from case to case.

The inadequacy of the universal treatment thesis, though, can be clear also for other evolutionistic considerations: evolutionism induces us to consider both the individual and the social consequences of individual psychopharmacological actions. Indeed, we need to consider also the possible dysgenic effects of the spread of psychopharmacological therapies. From the evolutionary point of view, one of the most important consequences of the efficacy of the pharmacological treatment of psychiatric disorders is that of equating the reproductive rates of the subjects affected by behavioral disorders with those of healthy individuals. Of course, these quotients tend to be different and to differ also as a function of the type of psychiatric pathology. For example, a depressive or anxious tendency can be associated with a comparatively low reproductive rate, while, conversely, a hyperthymic temperament can be expressed also in a high fertility rate.

In any case, any action exerted on the ailing phenotype tends to spread the genotype associated with the psychiatric disorder in the population and could thus lead to an increase of the disorder itself over time. In this connection, it would be interesting to make a study of the evolution of the incidence of certain psychiatric diseases by attempting to isolate the relationship between epidemiological trends and the number of treatments performed for these specific disorders.

CONCLUSION

The evolutionary approach emphasizes the individual dimension of diseases and, consequently, the need for personalized therapeutic actions. At the same time it suggests that the treatment should be geared to the achievement of the patients' objectives and to the maximization of their functional capacities inside a given context.

Drugs are prescribed on the basis of diagnoses, which normally follow the nosological criteria set out in DSM-IV and ICD-10. The classificatory logic on which this procedure is based does not correspond at all to the functional settings. The studies used to evaluate cohorts of normal or pathological individuals, in an attempt to render the response rates and prognostic evaluation homogeneous, prevent the identification of the individual variables that actually determine the standard deviations and, in most cases, the therapeutic or side effects due to psychotropic drugs. A greater knowledge of the recent progress made in Darwinian medicine could lead to a new therapeutic paradigm in which any necessary drug treatment must necessarily be integrated into the context and the life-style of the pathological individual.

Doing this is a moral requirement, given the twofold link between psychiatry and ethics spelled out in the second section. The flexibility of the Darwinian approach is needed, if psychiatry is to help in understanding what humans ought to do, and if psychiatric therapy has to help deficient human beings to be as they *can* and *ought* to be, without forcing untenable stereotypes on individuals.

REFERENCES

Ayala, F. 1972. The Autonomy of Biology as a Natural Science. In *Biology, History, and Natural Philosophy*, ed. A. Breck and W. Yourgrau, 1–16. New York: Plenum Press.
Baron-Cohen, S. (ed.). 1997. *The Maladapted Mind*. Hillsdale, N.J.: Lawrence A. Erlbaum Associates.
Bedau, M. A. 1992. Where's the Good in Teleology? *Philosophy and Phenomenological Research* 52: 781–806.
Bertilsson, L. 1995. Geographical/Interracial Differences in Polymorphic Drug Oxidation. Current State of Knowledge of Cytochromes P450 (CYP)2D6 and 2C19. *Clin. Pharmacokinet* 29: 192–209.
Brockmoller, J. 1999. Pharmacogenomics. Science Fiction Come True. *Int. J. Clin. Pharmacol. Ther.* 37: 317–318.
Brown, A. S., et al. 1998. Course of Acute Affective Disorders in a Developing Country Setting. *J. Nerv. Ment. Dis.* 186: 207–213.
Canali, S. (ed.). 2001. Drug-Abuse, Evolution, Medicine. *Medicina delle Tossicodipendenze – Italian Journal of Addiction. Monographic issue*: 9.

Canali, S., and Corbellini, G. K. R. (eds.). 2004. *Medicina darwiniana. L'approccio evoluzionistico alla malattia.* Bologna: Apèiron.

Corbellini, G. 1998. Le radici storico-critiche della medicina evoluzionistica. In *La medicina di Darwin,* ed. P. Donghi, 85–127. Bari and Rome: Laterza.

Damasio, A. R. 1994. *Descartes' Error: Emotion, Reason and the Human Brain.* New York: Putnam Books.

Davies, J. 1996. Origins and Evolution of Antibiotic Resistance. *Microbiologia* 12: 9–16.

Donghi, P. 1998. *La medicina di Darwin.* Bari and Rome: Laterza.

Foot, P. 2001. *Natural Goodness.* Oxford: Oxford University Press.

Goodwin, F. K., and Jamison, K. R. 1990. *Manic-Depressive Disease.* Oxford: Oxford University Press.

Hofbauer, K. G., and Huppertz, C. 2002. Pharmacotherapy and Evolution. *Trends in Ecology & Evolution* 17: 328–334.

Ingelman-Sundberg, M., Oscarson, M., and McLellan, R. A. 1999. Polymorphic Human Cytochrome P450 Enzymes: An Opportunity for Individualized Drug Treatment. *Trends Pharmacol. Sci.* 20: 342–349.

Kalow, W. 1991. Interethnic Variation of Drug Metabolism. *Trends Pharmacol. Sci.* 12: 102–107.

Kalow, W. 2001. Pharmacogenetics, Pharmacogenomics and Pharmacobiology. *Clin. Pharmacol. Ther.* 70: 1–4.

Knutson, B., et al. 1998. Selective Alteration of Personality and Social Behavior by Serotonergic Intervention. *Am. J. Psychiatry* 155: 373–379.

Kramer, P. D. 1997. *Listening to Prozac.* New York: Viking Penguin.

LeDoux, J. L. 1996. *The Emotional Brain.* New York: Simon and Schuster.

Levin, B. C., and Anderson, R. M. 1999. The Population of Anti-infective Chemotherapy and the Evolution of Drug Resistance: More Questions Than Answers. In *Evolution in Health and Disease,* ed. C. Stearns, 125-137. Oxford: Oxford University Press.

Lewis, A. J. 1934. Melancholia: A Clinical Survey of Depressive States. *J. Mental Sci.* 80: 1–43.

Marks, I. M., and Nesse, R. M. 1994. Fear and Fitness: An Evolutionary Analysis of Anxiety Disorders. *Ethol. Sociobiol.* 15: 247–61.

Masellis, M. T., et al. 2000. Pharmacogenetics of Antipsychotic Treatment: Lessons Learned from Clozapine. *Biol. Psychiatry* 47: 252–266.

McGuire, M. T., and Troisi, A. 1998. *Darwinian Psychiatry.* Oxford: Oxford University Press.

McGuire, M. T., Troisi, A., and Raleigh, M. M. 1997. Depression in Evolutionary Context. In *The Maladapted Mind,* ed. S. Baron-Cohen, 255–282. Hillsdale, N.J.: Lawrence A. Erlbaum Associates.

Murphy, M. C. 2001. *Natural Law and Practical Rationality.* Cambridge: Cambridge University Press.

Neel, J. V., Weder, A. B., and Julius, S. 1998. Type II Diabetes, Essential Hypertension, and Obesity as "Syndromes Of Impaired Genetic Homeostasis": The "Thrifty Genotype" Hypothesis Enters the 21st Century. *Perspectives in Biology and Medicine* 42: 44–74.

Nesse, R. M. 1999. Proximate and Evolutionary Studies of Anxiety, Stress and Depression: Synergy at the Interface. *Neuroscience and Biobehavioral Reviews* 23: 895–903.

Nesse, R. M. 2000. Is Depression an Adaptation? *Arch. Gen. Psychiatry* 57: 14–20.

Nesse, R. M., and Berridge, K. C. 1997. Psychoactive Drug Use in Evolutionary Perspective. *Science* 278: 63–66.

Nesse, R. M., and Williams, G. C. 1991. The Dawn of Darwinian Medicine. *Quarterly Review of Biology* 66: 1–22.

Nesse, R. M., and Williams, G. C. 1995. *Why We Get Sick*. New York: Random House.

Nesse, R. M., and Williams, G. C. 1999. Research Designs That Address Evolutionary Questions about Medical Disorders. In *Evolution in Health and Disease*, ed. C. Stearns, 16-23. Oxford: Oxford University Press.

Nesse, R. M., and Young, E. A. 2000. The Evolutionary Origins and Functions of the Stress Response. In *Encyclopedia of Stress*, ed. G. Fink, 79–84. San Diego: Academic Press.

Normark, B. H., and Normark, S. 2002. Evolution and Spread of Antibiotic Resistance. *J. Intern. Med.* 252: 91–106.

Ozdemir, V., et al. 2002. Pharmacogenomics and Personalized Therapeutics in Psychiatry. In *Neuropsychopharmacology: The Fifth Generation of Progress*, ed. K. L. Davis et al., 495-506. New York: Lippincott Williams & Wilkins.

Pani, L. 2000. Is There an Evolutionary Mismatch between the Normal Physiology of the Human Dopaminergic System and Current Environmental Conditions in Industrialized Countries? *Mol. Psychiatry* 5: 467–475.

Pincus, H. A., et al. 1998. Prescribing Trends in Psychotropic Medications: Primary Care, Psychiatry, and Other Medical Specialties. *JAMA* 279: 526–31.

Stearns, C. (ed.). 1999. *Evolution in Health and Disease*. Oxford: Oxford University Press.

Stevens, A., and Price, J. 1996. *Evolutionary Psychiatry*. London: Routledge.

Strawson, P. 1959. *Individuals*. London: Methuen.

Trevathan, W. R., Smith, E. O., and McKenna, J. J. (eds.). 1999. *Evolutionary Medicine*. Oxford: Oxford University Press.

Wiggins, D. 2001. *Sameness and Substance Renewed*. Cambridge:Cambridge University Press.

7

The Biology of Human Culture and Ethics

An Evolutionary Perspective

STEFANO PARMIGIANI, GABRIELE DE ANNA,
DANILO MAINARDI, AND PAOLA PALANZA

> I searched for great human beings; I always found the *apes* of their
> ideals.
>
> Nietzsche, *Götzen-Dämmerung*

> The human scientists proclaim that animals are irrelevant to the study of
> human beings and that there is no such thing as a universal human nature.
> The consequence is that science, so coldly successful at dissecting DNA,
> has proved spectacularly inept at tackling what the philosopher David
> Hume called the greater question of all: why is human nature what it is?
>
> Ridley 1993

The traditional nature-nurture controversy concerns whether ethical norms
and social practices depend on cultural influences on individuals and societies
or whether they are based on the particular features of our nature. In this
chapter, we argue that nature and nurture cannot be sharply distinguished, as
some scientists have tried to do in the past. Evolutionary considerations lead
us to conclude that facts about our nature and the ways in which we come
to form cultural traditions are entrenched in evolutionary processes. Before
turning to the issue, though, we thought it might be worthwhile to deal first
with two possible misunderstandings.

First, when we claim that we are concerned with the "bases" of ethical
norms and practices, we do not mean that we offer a precise account of the
ways in which such bases are relevant for the explanation or the justification of
ethics. In particular, we do not deal with the question whether ethical systems

We wish to thank Professor Elizabeth Ferrero, St. Thomas University, Miami, Florida, for encour-
agement and advice and Professor G. Boniolo, University of Padova, for advice and discussions.

121

or moral norms can be reduced to natural facts or cultural facts. We claim that some natural and cultural facts are ethically relevant, but we do not explain whether they are relevant because ethical facts can be reduced to them, or because they can be the truth makers of ethical claims, or because they can be the premises of arguments leading to ethical conclusions or whatever one may think the relevance of these facts for ethics may be. That these facts are relevant for ethics seems to us to be an intuitive truth, and the relevance of each can be compared with that of another, even if one has not previously specified in what way they are relevant. However, we do have something to say that might be interesting to those concerned with reductionism. Very often, people take the nature position in the nature-nurture debate to lead toward reductionism, because that stand is taken to amount to genetic determinism. We claim, however, that the nurture position is no less inclined to determinism and that an evolutionary perspective opens up several doubts about genetic determinism.

Second, when we speak of "human nature" we do not refer to any particular philosophical conception of man, or to any individual perspective on our species coming from a particular science. For example, we do not think about an Aristotelian, Marxist, or Hegelian conception of man. Nor do we think about what cognitive psychology or anthropology or genetics can teach us about our species. Rather, we leave it as an open question what the characteristics of our species can be, and we seek a unifying conception by joining together data from different disciplines. Borrowing Wittgenstein's idea of the logical space as the set of all possible combinations of objects, we can understand "human nature" as a logical space of phenotypical traits (biological, psychological, behavioral, etc.), which is a function of genotypes and environmental variables. Evolutionary considerations help us to determine this logical space, that is, to enlarge and refine our conception of our nature. As a consequence, we can claim that certain ethically relevant facts are universal features of our species and not the mere outcome of culture.

In the next section, we sketch the nurture-nature debate and underline some aspects of it that are relevant for our discussion. In the following sections, we consider the importance of the Darwinian approach for the study of behavior, both from an ethological and a psychological point of view. Next we consider the case of sexual behavior from a Darwinian perspective; that case represents a fundamental contribution of the Darwinian approach and is also a paradigmatic example of ethically relevant behavior, which is strongly based on biological truths about our species. Then we offer some general remarks concerning the relevance for ethics of the aforementioned biological facts.

THE NATURE-NURTURE DEBATE

The nature-nurture debate can be framed within a traditional controversy between two different approaches to human behavior: ethology, on the nature side, and behavioral and comparative psychology, on the nurture side.

Ethology (or biology of behavior or comparative analysis of behavior, as K. Lorenz, one of its founders, used to call it) is based on a comparative analysis of stereotypical instinctual patterns of behavior (known as fixed action patterns), across closely related species. Through these analyses, ethologists came to the conclusion that several kinds of behavior are largely inherited (i.e., genetically determined) and that they are very poorly influenced by individual experiences. Like body traits (e.g., the colors and shape of the tails of peacocks), these kinds of behavior were considered species-specific characteristics and the results of natural selection.

For several comparative psychologists, instead, genes have nothing to do with behavior, which has to be regarded as a product of learning. These positions have their roots in John Locke's philosophical idea, according to which, at birth, the human mind would be a *tabula rasa*, an "empty board." According to him, it is only through experience (i.e., through learning and through the social and cultural environment) that the mind of higher animals (i.e., mammals), and especially the human mind, can be filled. Thus, human nature would be an expression and a development of cognitive and social experiences.

Roughly, this is also the theoretical background of the American psychological mainstream known as "behaviorism" (Cartwright 2000). The following quotation from Watson, who, as it is well known, was a leading behaviorist, expresses neatly the behaviorist position: "Give me a dozen of healthy infants and I'll guarantee to take any one at random and train (i.e., provide him with the right *experiences, social environment and learning*) him to become any type of specialist. I might select – doctor, lawyer, and even beggar and thief, regardless of his talent, tendencies, vocations, and race of his ancestors" (Watson 1930).

In recent times, however, the nurture-nature dichotomy has been considered a false problem, because the controversy was diminished by the contribution of modern genetics. The solution was offered by the concept of phenotype, which refers to the result of the expression of genes and of their interaction with developmental and environmental factors. A phenotype may be a protein, or a brain structure, or even a kind of behavior (cf. Boniolo and Vezzoni, Chapter 5 in this volume). Behavior, then, has a genetic basis, and thus it is impossible to claim that behavior has a purely learned, cultural, or social base. Even the more complex kinds of behavior have a genetic base. For example, one's learning capacities depend on one's genes. The ability to learn and to use language is "genetic," in the sense that genetic instructions cause (contribute to) the constitution of a specific human brain, which includes a certain language-acquisition device.

Complex kinds of behavior, however, are not purely "natural," because they are phenotypical traits, the realization of which requires also the attainment of suitable environmental conditions. These may also include social and cultural conditions. Let us consider the example of language again. What languages a human being speaks is a contingent matter, depending on one's country and sociocultural environment (Ridley 1993).

The upshot of the new perspective is that no kind of behavior can be said to be purely inherited or purely learned. It is always the result of a complex combination of genetic and environmental factors. The latter may include cultural and social aspects. Thus, when contemporary scientists speak of a behavior being species-specific and genetically based (or even *strongly* genetically based), they do not mean that it is *fully* determined by the genes. It may be worth considering a simple example: courtship in male mammals.

A male courting behavior in a mammal species is always the result of several cascade events, both biological and nonbiological. The zygote must be XY in order to be a genetically potential male; the early formed embryonic gonads must deliver testosterone in the blood stream; in turn, the sexual steroid hormone must stimulate target cells (which must have already developed the proper genetically determined protein receptors), both at the level of genital organs and in the brain. This is a necessary condition for the masculinization of the developing individual. If all of these events happen and the individual survives long enough, at puberty, through the action of the male hormone (and so, ultimately, through the actions of the male genes in the Y chromosome, which contains the "program" needed to develop male gonads with an encoded genetic information suitable to produce the enzymes that may convert cholesterol into testosterone), the male individual will be able to react to specific sexual stimuli (i.e., pheromones, bright colors, behavior), which are emitted by a co-specific female. In order for the courting behavior to take place, however, the male individual must also have developed an appropriate body coordination, which depends on its cognitive development. The latter may eventually be "learned" from the cultural and social environment, for example, from its parents.

Courting is a kind of behavior that is normally considered strongly based on genetic inheritance. It is clear from the example, however, that when scientists say that, they do not mean to deny that there may be other nongenetic conditions. Probably, several disagreements among philosophers, human or social scientists, and evolutionary biologists derive from a misunderstanding of this point. For a social anthropologist who observes the malleability of human behavior, for example, the idea that mind processes may be genetically modulated might at first seem an absurdity. A deeper understanding of genetic developmental biology, however, would probably lead to a quite different conclusion.

DARWIN'S THEORY OF EVOLUTION AND THE UNDERSTANDING OF OUR NATURE

We can now consider how an evolutionary approach can help to explain human phenotypical traits, including ethically relevant sorts of behavior.

It must be stressed that the evolutionary approach can hardly be doubted. In the scientific community, at present, there is no doubt that life and its biodiversity originated through the process of evolution by natural selection (Darwin 1859). Even if Darwin had not offered his tremendous contribution to science, biologists from different disciplines would still recognize that the diversity of life is based upon an essential "basic unity" (e.g., the universality of the genetic code). Molecular biology, biochemistry, genetics, molecular genetics, comparative anatomy, comparative embryology, developmental biology, taxonomy, ethology or/and comparative psychology, neuroscience, and so forth, all point to that. Thanks to the study of microevolutionary processes (i.e., the changes of the frequency of the variations of the same gene, the so-called alleles), evolution has been experimentally tested. Fossils are probably the best evidence for the claim that species originated from common ancestors (i.e., macroevolutionary processes). But strong evidence comes also from molecular evolution (i.e., the fact that the molecular analysis of proteins and more-specific DNA sequences in phylogenetically related species highlights homologous genes and proteins). In this respect, our closely related species are chimpanzees (*Pan troglodytes*) and bonobos (*Pan paniscus*), with which we share almost 98 percent of the genes.

A widespread opinion, common even among well-educated people, takes the scientific foundation of evolutionism to be merely hypothetical. It must be noted, though, that, contrary to that opinion, evolutionary biologists discuss or disagree about the mechanisms of macroevolutionary processes (e.g., are natural selection and genetic mutation the only mechanisms involved?), but evolution itself is not in question.

125

The current account of how evolution takes place by means of natural selection, which is known as neo-Darwinism or the modern synthesis, can be summarized as follows: natural selection is fundamentally a mechanism that acts *on genetically based phenotypical* variations (i.e., morphology, physiology, and behavior) of *individuals* within a population of a species living in a given environmental situation and that results in the *differential survival and reproduction (fitness) of individuals.* In his book *On the Origin of Species,* Darwin was very reluctant to apply his idea to humans, since the consequences of his views were very clear; this led to a very hot debate, not only among scientists but also among the general public. Even now, more or less consciously, people generally accept that the "body" of humans could have originated through evolution, although they feel that human behavior, and hence the human mind, has nothing to do with evolution. It must be the result of human "free will." Thus, they say, it must be a product of culture, not of the genes. Otherwise, it would be fully determined. Several social scientists and philosophers still endorse this line of thinking, as if the revolution of evolutionism had never taken place (Ridley 1993).

The implication of the theory of natural selection for the evolution of behavior was scientifically addressed by Darwin in his books *The Descent of Man and Selection in Relation to Sex* (1871) and the *Expression of the Emotions in Man and Animals* (1872). The basic assumption of Darwin's view is that humans and animals share a phylogenetic continuity in body structure and in the mind (i.e., human and animal minds differ only in degree and not in kind). In his view, the "miracle" (i.e., the complexity and uniqueness) of the human mind is an emergent property of the evolutionary processes that shaped the biological organization of the brain. Darwin's intuition about the possibility to apply his theory to the study of psychology is well expressed in this claim: "I see open fields for more important researches. Psychology will be based on new foundations, that of the necessary acquirement of each mental power and capacity by gradation. Light will be thrown on the origin of man and his history" (Darwin 1859). Unfortunately, most of Darwin's writings on the expression of emotion were heavily anecdotal and anthropomorphic, due to the paucity of experimental data available at his time. Consequently, for almost another hundred years, human and social scientists kept considering social norms and ethics as a mere product of culture. In the twentieth century, however, important contributions to the study of behavior within a Darwinian framework were introduced by the development of ethology, sociobiology (i.e., the study of the evolution of animal and human social behavior), and evolutionary psychology (which we prefer to call bioevolutionary psychology).

The main tenet of the evolutionary approach to the study of behavior is the idea that behavior cannot be fully explained by cultural and social conditions and that biology has to play a role. The new attitude is well expressed by John Tooby and Leda Cosmides (1990): "The assertion that 'culture' explains human variation will be taken seriously when there are reports of women war parties raiding villages to capture men as husbands, or parents cloistering their sons but not their daughters to protect their sons' virtue, or when cultural distributions for preferences concerning physical attractiveness, earning power, relative age and so on show as many cultures with bias in one direction as in the other."

The evolutionary approach to the study of behavior radically changed the way in which biologists ask questions about the behavioral phenotypes. After the Darwinian revolution, a scientist is not merely interested in the mechanisms underlying a certain behavior (i.e., how a behavior comes to be expressed): the most intriguing and important question now is why a certain behavior comes to be expressed (i.e., the adaptive significance, alias the evolutionary reason). Behavioral biologists maintain that a rigorous theoretical conceptualization of behavior raises two different questions: the question *how* and the question *why* a certain behavior comes to be expressed. We can distinguish between proximal mechanisms or proximate causes of behavior, which encompass all sorts of factors (e.g., genes, development, physiology, experience) that explain how or in which way the behavior is expressed; and ultimate causes or ultimate mechanisms of behavior, which explain why a given behavior and its proximal mechanisms have been favored by natural selection. The question *why* explains its adaptive function and thus the value for survival of the underlying mechanisms (e.g., neurochemical substrates). From this perspective, context and function are of paramount importance for the study of behavior.

Laboratory ethologists do not constrain their observations on animals only to the study, for example, of general rules concerning learning and memory, or of the laws of aggressive behavior. They also try to simulate the natural situations in which the behavior has the highest probability to be expressed. In this way, they can gain a better understanding of proximal mechanisms, their evolution and their adaptive value. For example, we can correctly say that one of the proximal mechanisms promoting or evoking intrasexual aggression between a resident territorial male mouse and a co-sexual co-specific intruder are the pheromonal cues coming from the intruder and the level of testosterone in the resident. But the evolutionary perspective opens a further question, concerning what we called the ultimate causes: why do males attack and become aggressive toward co-specifics of the same sex? An answer may be that they compete for the reproductive opportunity. Indeed, as we go on to explain, the real Darwinian revolution was not the definition of natural selection (which had been suggested also by Alfred Wallace), but the idea that most of the dimorphisms in body structure, and in behavior (and, hence, in the mind) between the two sexes is due to sexual selection (Cronin 1992; Miller 1992, 2000).

DARWINISM AND SEXUAL SELECTION THEORY

Darwin thought that natural selection acts mainly on the individual phenotype (and, of course, indirectly on the genes responsible for its expression). This notwithstanding, most evolutionary biologists (including ethologists) were convinced that natural selection promotes the so-called benefit of the species. For a long time, the conceptual and philosophical orthodoxy was that any character increasing the survival of the species would be favored by natural selection. This idea was strongly advocated, for example, by the eminent evolutionary behavioral biologist K. Lorenz, the father of ethology. In his book *On Aggression* (1966), he claimed that animals evolved an inhibition to kill members of the same species for the "good of the species." This thesis is better known as "group selection theory." The core of natural selection is reproduction. In fact, only the individuals who are successfully reproducing contribute to the "genetic pool" of the successive generation and, hence, to the species. In other words, the genes and phenotypes (including behaviors) of a given species are those of the individuals that survived long enough to reproduce. With this in mind, we can rephrase the evolutionary slogan according to which selection operates only through the survival of the fittest, in terms of the reproduction of the fittest (Ridley 1993).

The evolutionary biologist G. Williams (1966), however, pointed out that the "group selection theory" is fallacious, at least from a Darwinian perspective. If individual reproduction is the most important thing in terms of genes transmission, it turns out that in sexually reproducing species there is a strong intraspecific competition for reproduction. Darwin had deep insights on this issue. Indeed, he was impressed by the dimorphism between the two sexes (e.g., by the fact that males have weapons to fight rivals in order to achieve territory and mates; or by the fact that they have ornaments, e.g., the peacock tail). This dimorphism could not be satisfactorily explained by the action of natural selection, simply intended as external factors acting on individuals, like environmental factors (e.g., climatic variations) or predators. In fact, many ornaments (bright colors, long tails, or heavy antlers) or elaborate fighting and courtship behavior are costly in energetic terms and can attract predators. They can only find their adaptive significance in the light of Darwinian sexual selection theory (Darwin 1871). In synthesis, Darwin claimed (as we now commonly believe) that a major force in the evolution of certain bodily and behavioral traits of a species is operated by co-specifics through competition for the opportunity to mate and reproduce. We can distinguish two cases of sexual selection: intrasexual selection and intersexual selection. The antlers and many weapons that we see in males represent the adaptive outcome of the competition for mating, expressed as intrasexual aggression, and are examples of intrasexual selection. On the other hand, the bright colors and the conspicuous peacock tail are the result of mate choice operated by the female. These are clear examples of intersexual selection.

The rediscovery of sexual selection theory and thus of the importance of individual reproductive advantage (i.e., the individual fitness) made biologists fully understand the power and logic of natural and sexual selection for the evolution of behavior. Sexual selection maximizes the individual fitness and the probability to pass its genes to the next generation. From this point of view, natural selection and sexual selection are the differential reproductive success of individuals at the expense of co-specific members. This is telling about the nonequalitarian "logic" of nature. This kind of reasoning led – in the second half of the twentieth century – to the birth of sociobiology, started by E. O. Wilson (1975) and radically expressed by R. Dawkins in his book *The Selfish Gene* (1976). Essentially, the individual is considered as a "machine" that replicates genes. In this perspective, the self-interest of the individual in terms of reproductive fitness is the interest of its genes (i.e., genes that contain a program to develop a phenotype that may be successful in the competition with co-specifics for reproductive opportunities).

The superiority of the selfish-gene theory over the group selection theory has subsequently been supported by several cases. One of the clearest of these is "sexually selected infanticide" (Hrdy 1974). Infanticide was first observed in a species of langur (*Presbitis entellus*) and in lions (*Panthera leo*). Now, it is well documented in numerous mammalian species, vertebrates and invertebrates (Parmigiani and vom Saal 1994). Usually, infanticide is performed by males (but eventually also by females) when taking over a reproductive territory from a co-specific of the same sex. The usurper kills the offspring of the competitor, in order to get sexual access to the female. In the case of langurs, for example, the female cannot be inseminated by the usurper before it resumes ovulation after the end of lactation. Thus, killing the offspring of the previous male accelerates the reproductive process of the usurper.

Hrdy suggested that this behavior is not pathological but is adaptive: the evolutionary force or ultimate cause of this behavior is intrasexual competition for mating opportunities. Sexual selection must thus be responsible for the origin of the mechanism underlying this type of infanticide. The theoretical tenets of this hypothesis are that there must be a reproductive advantage for the infanticidal individual; the behavior must have, at least in part, a genetic base (i.e., without genes, selection cannot work and there is no evolutionary change); and there must be an inhibition to kill one's own offspring. In the laboratory, observations on mice confirmed the theoretical model. They also showed that intrasexual aggression and infanticide share a similar neurochemical substrate (Parmigiani and Palanza 1991; Parmigiani et al. 1998).

The example of infanticide shows the "amorality" of nature. In fact, a behavior that causes harm or killing of members of the same species (i.e., a behavior that we would call immoral when exhibited by a member of our species) enhances the propagation of the individual genes (i.e., individual fitness), and thus natural selection and sexual selection spread those genes into the population. We can then ask ourselves why we consider immoral that sort of behavior, that is, why we think that infanticide should not be practiced. There are two possible answers, from an evolutionary perspective. Either our species has evolved in a way such that infanticide does not improve the chances of individual reproduction. Or the evolution of our species has promoted the emergence of a moral mental capacity that leads us to behave in ways that do not maximize the chances of individual reproduction. In the first case, ethics would follow the principles of evolution. In the latter case, there are two possibilities. Either the evolution of our species does not follow the mechanisms of the selfish-gene theory (and an alternative mechanism of evolution must be at work in our case, e.g., that described by group selection theory) or the evolution of our species through the pattern described by the selfish-gene theory led to a form of life that does not follow that very mechanism of evolution. The point, then, is: where do our moral capacity and our ethical systems come from. Do they evolve?

Although man can be defined as a "cultural and moral animal," the application of the evolutionary approach to human behavior is based on the assumption that, as in the case of other animals, natural selection and sexual selection have "forged" human nature. From this perspective, it is legitimate to seek an understanding of human nature through the comparative methods used in classical ethology. It is also possible to apply the evolutionary logic in order to make predictions about human behavior and the ultimate causes of his mental capacities, including the moral capacity.

Eibl-Eibesfeldt, one of Lorenz's disciples, founded human ethology precisely for the purpose of addressing these questions (see Eibl-Eibesfeldt 1989). He studied the expression of emotions in people from different races and cultures and clearly showed that facial expression and certain patterns of behavior (e.g., in contexts such as mother-child interactions or courtship) are human universals. They are species-specific characteristics. Social and human scientists, on the other hand, may counter that these are just examples from a very small group of genetically based kinds of human behavior. Complex kinds of behavior (e.g., social relationships like the choice of sexual partners and marriage) and normatively guided conducts, the objection may go, are so culturally diverse across humankind that human behavior must be influenced by genes to a very small degree. Learning and culture play the central role.

A reply to this objection may come from a new way to approach the study of human psychology within the evolutionary framework, namely evolutionary psychology (Buss 2001). The assumption of this approach is that human psychology and human societies (including social rules and ethics) are not the mere result of culture. Rather, cultures, ethical norms, and social practices are the result of special capacities of the human mind, which originated through evolution. If this assumption is correct and, thus, our conducts originated from evolution, like those of other species, we might expect human nature to be universal. At least a certain number of kinds of behavior, for example, those related to sexual selection and intraspecific competition, must manifest patterns similar to those – considered previously – which may be observed in other species, when we apply the same ethological paradigm.

In the terms of ultimate causes of behavior related to reproductive strategies, we should succeed in making predictions on differences among the sexes, concerning jealousy, mate selection, and abuse of the young in our species. The evolutionary psychologists maintain that, if there is a human mental capacity that evolved through selection and is a universal characteristic of human nature, we can predict that differences among the sexes concerning the previously mentioned kinds of behavior must be minimal across different human cultures.

Because, as in other mammals, human females have a greater parental investment than males, we expect that females are very selective and look for males with clear evidence of health (Ridley 1993; Buss 1992, 1999); furthermore, the power of sexual selection, which depends mainly on choices by the females, must have influenced the evolution of human behavior and – possibly – of mental capabilities (Cronin 1992; Miller 1992, 2000). In a study on mate preferences involving more than 10,000 people from different cultures and religions in thirty-seven countries, the evolutionary psychologist

D. Buss (1989) showed that males are more interested in the youth, beauty, intelligence, and kindness of their partners. Women, instead, are concerned about the kindness, sensitivity, intelligence, wealth, and social status of their partners. Men value chastity (sexual loyalty) significantly more highly than women. Interestingly enough, there are no cultures in which the reverse is true (see Tooby and Cosmides 1999). Thus, beyond cultural traditions (religious and/or nuptial systems), there is a universal human characteristic: men look at "indicators of fertility" (youth and beauty) and sexual fidelity (i.e., certainty of paternity), whereas women pay more attention to the social status (power) and wealth of males. In an evolutionary perspective, women look for good genes and resources.

It must be noted that, contrary to what might be the case at present, cognitive capacities and the skills needed to achieve resources were strongly interlinked in our ancestors: the most intelligent individuals were most likely to be those who could provide the best or most resources. Because this trait is genetically based (cf. Zechner et al. 2001), these men manifested to women also good indicators of their good genes (e.g., verbal skill, wittiness, creativity, a tendency to get the leadership in a group of hunters). As Cronin (1992) pointed out, the power of female choice on the peculiar human trait that we call intelligence must have been crucial in the evolution of our mental capacities.

We can apply the same evolutionary logic to test hypotheses about the ultimate causes of infant abuse and sexual jealousy. From an evolutionary point of view, male sexual jealousy can serve the function of avoiding wasting one's efforts and energy in parental caring for children sired by other men. After all, the mother has always been certain, whereas paternity was only probable, before some recent advances in genetic engineering. If jealousy related behaviors have a genetic basis, these genes have been maintained in the human genome. Evolutionary psychologists Margo Wilson and Martin Daly (1994) studied jealousy-related behavior in humans and concluded that it agrees perfectly with an evolutionary interpretation. Jealousy seems to suggest that human evolution followed the mechanism of selection described by the selfish-gene theory. In fact, they showed that different human societies, despite cultural diversities, exhibit "monotonically similar" ethical rules, such as the social recognition of marriage, the condemnation of adultery as a violation of property, the value of female chastity and female "virtue," and the control of female sexuality. In most societies, it is maintained that females have to be protected by their males, and, in this way, by means of specific social rules, men try to prohibit sexual contact by their females with other males. Thus, it seems that jealousy is a universal "dangerous passion" (Buss 2001) of human nature (cf. also Ridley 1993).

This strongly suggests the hypothesis that, like lions, langurs, mice, and apes (e.g., chimpanzees), humans have the tendency to discriminate between parental care given to their own children and that given to adopted children. Wilson and Daly (1994) report data that confirm that abuse and/or infanticide are significantly more frequent when the children are not the biological offspring of their parents.

THE BIOLOGICAL ROOTS OF SOCIAL NORMS AND ETHICS

The foregoing considerations about the neo-Darwinian framework involving sexual selection and the evolution of human mental traits allow us to speculate about the biological roots of certain ethical norms, social prescriptions, and even divine commandments. The ethologist W. Wickler (1971) explored the possibility of a continuum between naturally and culturally determined human complex social behavior in his book *Die Biologie der Zehn Gebote* (*The Biology of the Ten Commandments*). For example, the commandments of the Bible forbid desiring the resources and "the woman of another man" (Exodus 20:17). Obviously, women are here considered the property of men, and this shows that when these ethical rules were written, an affair with the wife of another man was considered a violation of property.

Other examples taken from the Bible are even more astonishing, if they are reinterpreted in the light of the previously considered evolutionary explanation of competitive aggression and sexual selection. When "his elected people" were entering a foreign country in order to conquer new resources, God commanded that they had to "kill every male even the young, and kill every woman that had sexual contact with males. But all young women, that have not known a man, keep alive for yourselves" (Numbers 31:17). This is surprisingly similar to the case of sexually selected infanticide in other mammals. And it seems equally humanely and morally unacceptable to our moral sensitivity. Interestingly, God commanded also that a foreign culture should be destroyed, in order to keep the belief and related norms of the "people of God" uncontaminated (Deuteronomy 20:18). In this connection, it must be recalled that human beings do not transmit only "genetic information" to the next generation but also "cultural information" through the process of learning and teaching. "Memes," the term Dawkins (1976) uses to define these "pieces of cultural information," are needed to account for the continuum between biological evolution and cultural evolution, which are considered parts of the same process. Biology and culture interact and play the game of the evolution of human nature. In fact, gene variations influence culturally determined kinds of behavior, and, the other way around, variations in culture and traditions (i.e., the "mutant meme") provide as feedback selective pressure on genes.

In conclusion, it seems that, possibly among other factors, sexual selection played a role in shaping the human mind and human behavior: if this is true we cannot fully understand our nature without understanding how it evolved. Thus, we need to understand the origins of sexual reproduction, of related behavior, and of sexual selection mechanisms (cf. Ridley 1993). In the past, many hypotheses have been proposed concerning the evolution of our astounding mental capacities, including reason and, in particular, self-consciousness. Among them, we recall the thesis of a very rapid feedback between biogenetic and cultural evolution, involved in the use and manufacturing of tools. Another one is the idea of a cooperation between protohuman hunters. Building on an intersexual selection theory (involving the coevolution of male characteristics and behavior, on the one hand, and female preferences for those characteristics, on the other), the evolutionary psychologist Geoffrey Miller proposed that our astounding brain capacities are the result of a sort of runaway sexual selection:

> I suggest that the neocortex is not primarily or exclusively a device for tool making, warfare, hunting or avoiding savanna predators. None of the postulated functions alone can explain its explosive development in our lineage and not in closely related species. The neocortex is largely a courtship device to attract and retain sexual mates: its specific evolutionary function (*ultimate cause*) is to stimulate and entertain other people, and to assess the stimulation attempts of others. (1992)

Indeed, in relationships among animals (e.g., in competitive aggressive interactions), the ability to understand the co-specifics' emotions and predict their behavior (which is a very important skill in a highly social animal, like a primate) evolved *at least in part* under the pressure of intrasexual selection. The application of the game theory models indirectly revealed that animals use an emotional calculus to achieve information about the intentions of co-specifics: when A and B compete, the behavioral strategy of A and thus the outcome of fighting depends on what B does. Obviously, the capacity to lie is evolutionary advantageous in a game based on individual advantage or – in evolutionary terms – on the "selfish gene." The same thing applies to courtship behavior, as a form of intersexual selection (Dawkins 1976).

The majority of human courtship is based on verbal skills, wittiness, creativity, and long talk. During these interactions a man and a woman can get information about the quality (kindness and intelligence) of the prospective mate, with whom they might share the adventure of reproduction. Are these kinds of behavior and mental capabilities indicators of fitness and good genes? According to this hypothesis, our "big brain" with its power of awareness and consciousness has been "designed" by sexual selection to understand "what's going on in the mind of other people" and especially in sexual relationships. In this process, the brain became capable "to see and perceive itself" and, eventually, became capable of cognition and conceptualization of the underlying emotional states.

Recent observations of an excess (compared with autosomes) of genes responsible for cognitive abilities in the X human chromosome (which, among other things, affects fertility) implies that this character is selected for and this tends to support Miller's hypothesis (2000) that the human mind (i.e., cognition) has been shaped during evolution by female choice (Zechner et al. 2001).

CONCLUSION

At this point, it seems to us that we can explain why, as we stated at the beginning of this chapter, one can intuitively claim that science provides the bases for ethics, although one does not know how to (or does not want to) explain how those bases are relevant. In fact, as we have seen, evolutionary psychology suggests that psychological traits like jealousy, attitudes toward infanticide, and preferences about the properties of mates are universal characteristics of human nature. These attitudes, on the other hand, concern kinds of behavior that are the objects of normative assessments. Indeed, virtually all ethical or legal systems that we might encounter deal – maybe in divergent ways – with these sorts of behavior. At the same time, however, the data that we have pointed to and the arguments that we have suggested do not entail one particular way in which science can help explain or justify ethical systems.

Our discussion, on the other hand, seems to suggest two important points to one who attempts to explain or justify ethical systems.

First, appropriate explanations and justifications must consider the fact that ethics is not the mere result of culture, because ethics concerns at least some kinds of behavior that depend on nature: the mentioned psychological attitudes, indeed, are universal features of human nature. We have seen that we can speculate that this is true also of higher cognitive and rational capacities, which may be relevant for the development of ethical systems and for normatively guided types of life. In other words, it is plausible that our moral capacities are universal features of human nature, which were selected throughout evolution.

As we noted, the preference for the cultural horn in the nature-nurture controversy usually originates from the desire to avoid determinism, and thus save ethics from a collapse due to the denial of free will and the consequent refutation of responsibility. We can now see, however, that this preoccupation is misplaced, for at least two reasons. First, we can now understand why the fact that moral capacities and ethical systems depend on genetic inheritance does not imply that ethics is determined by genetics. As we have seen, genes are just one among several types of causes responsible for our ethical pheno- type. Other types of causes must also be considered, and it remains an open question whether any sufficient set of causes can be deterministically speci- fied. Were determinism false, furthermore, it would also be an open question whether indeterminism is only epistemic or goes all the way down to the onto- logical level. Second, cultural explanations of ethics are no less dangerous for free will than those based on genetics. Cultures, in fact, may be claimed to fully determine the psychological and volitional antecedents of action, with the result that there is no room for free will. "Nurtural" determinism is no less dangerous for ethics than "natural" determinism.

The second point relevant for those who attempt to explain and justify ethics through biology is the following. Even if our arguments do not suggest any particular indication on how biological facts are relevant for ethics, they do put some interesting constraints on viable possible explanations or justifications of ethics through biology. Indeed, the evolutionary approach to man and its cultural products (i.e., morality, ethics, and even religion), far from being a dogmatic biological determinism, may contribute, through the interactions with other disciplines (e.g., human and social sciences, philosophy, and also theology), to understanding the human uniqueness and dignity. The previously considered examples indicate that our human biological nature and human cultures cannot be sharply separated, but they constitute a continuum. It is true that we presently lack a suitable theory that may succeed in integrating cultural evolution with biological evolution. Consequently, biological and social sciences lack a unifying explanatory schema that may account for the origin and evolution of culture. But the upshot of the examples we have considered here is that the evolutionary approach to human nature may provide a unified theoretical background. The biological genetically based evolution of the human species and the Lamarckian evolution of culture can and must be merged in an integrated explanation. This purports that all attempts to explain or justify ethics that keep human nature and human culture sharply distinct must be fundamentally flawed.

REFERENCES

Buss, D. 1989. Sex Differences in Human Mate Preferences: Evolutionary Hypothesis Testing in 37 Cultures. *Behavioral and Brain Sciences* 12: 1–49.

Buss, D. 1992. Preference Mechanisms in Human Mating: Implications for Mate Choice and Intrasexual Competition. In *The Adapted Mind*, ed. J. Barkow, L. Cosmides, and J. Tooby, 249–266. New York: Oxford University Press.

Buss, D. 1999. *Evolutionary Psychology: The New Science of the Mind*. Boston: Allyn and Bacon.

Buss, D. 2001. *The Dangerous Passion*. London: Bloomsbury Publishing.

Cartwright, J. 2000. *Evolution and Human Behavior*. London: Macmillan.

Cavalli Sforza, L. L., and Feldman, M. 1981. *Cultural Transmission and Evolution*. Princeton: Princeton University Press.

Cronin, H.1992. *The Ant and the Peacock*. Cambridge: Cambridge University Press.

Darwin, C. 1859. *On the Origin of Species*. London: Murray.

Darwin, C.1871. *The Descent of Man and Selection in Relation to Sex*. London: Murray.

Darwin, C.1872. *Expression of the Emotions in Man and Animals*. London: Murray.

Dawkins, R. 1976. *The Selfish Gene*. Oxford: Oxford University Press.

Eibl-Eibesfeldt, I. 1989. *Human Ethology*. Hawthorne, N.Y.: Aldine de Gruyter.

Hrdy, S. 1974. Infanticide among Animals: A Review, Classification and Examination of the Implication for the Reproductive Strategies of Females. *Ethology and Sociobiology* 1: 13–40.

Lorenz, K. 1966. *On Aggression*. New York: Harcourt Brace.

Mainardi, D. 2001. *L'animale irrazionale*. Milan: Mondadori.

Miller, G. 1992. Sexual Selection for Protean Expressiveness: A New Model of Hominid Encephalization. Paper presented at the 4th annual meeting of the Human Behavior and Evolution Society, Albuquerque, New Mexico.

Miller, G. 2000. *The Mating Mind: How Sexual Choice Shaped the Evolution of Human Nature*. London: Vintage.

Parmigiani, S., Ferrari, P. F., and Palanza, P. 1998. An Evolutionary Approach to Behavioral Pharmacology: Using Drugs to Understand Proximate and Ultimate Mechanisms of Different Forms of Aggression in Mice. *Neuroscience and Biobehavioral Reviews* 23: 143–153.

Parmigiani, S., and Palanza, P. 1991. Fluprazine Inhibits Intermale Attack and Infanticide, but Not Predation in Male Mice. *Neuroscience and Neurobehavioral Reviews* 15: 511–551.

Parmigiani, S., and vom Saal, F. 1994. *Infanticide and Parental Care*. Chur, Switzerland: Harwood Academic Publisher.

Ridley, M. 1993. *The Red Queen: Sex and the Evolution of Human Nature*. Viking: London.

Tooby, J., and Cosmides, L. 1990. On the Universality of Human Nature and the Uniqueness of the Individual: The Role of Genetics and Adaptation. *Journal of Personality* 57: 17–67.

Watson, J. 1930. *Behaviorism*. New York: W. W. Norton.

Wickler, W. 1971. *Die Biologie der zehn Gebote*. Munich: Piper.

Williams, G. C. 1966. *Adaptation and Natural Selection*. Princeton: Princeton University Press.

Wilson, E. O. 1975. *Sociobiology: The New Synthesis*. Cambridge, Mass.: Harvard University Press.

Wilson, M., and Daly, M. 1994. The Psychology of Parenting in Evolutionary Perspective and the Case of Human Filicide. In *Infanticide and Parental Care,* ed. S. Parmigiani and F. vom Saal, 73–105. Chur, Switzerland: Harwood Academic Publisher.

Zechner, U., et al. 2001. A High Density of X-Linked Genes for Cognitive Ability: A Run-Away Process Shaping Human Evolution? *Trends in Genetics* 17: 697–701.

IV

How Biological Results Can Help
Explain Moral Systems

8

Biology to Ethics

An Evolutionist's View of Human Nature

FRANCISCO AYALA

Humans are animals but a very distinct and unique kind of animal. Our anatomical differences include bipedal gait and enormous brains. But we are notably different also, especially in our individual and social behaviors and in the products of those behaviors. With the advent of humankind, biological evolution transcended itself and ushered in cultural evolution, a more rapid and effective mode of evolution than the biological mode. Products of cultural evolution include science and technology; complex social and political institutions; religious and ethical traditions; language, literature, and art; and electronic communication.

In this chapter, I explore ethics and ethical behavior as a model case to illuminate the interplay between biology and culture. I propose that our exalted intelligence – a product of biological evolution – predisposes us to form ethical judgments, that is, to evaluate actions as either good or evil. I further argue that the moral codes that guide our ethical behavior transcend biology in that they are not biologically determined; rather, they are products of human history, including social and religious traditions.

HUMAN ORIGINS

Mankind is a biological species that has evolved from species that were not human. Our closest biological relatives are the great apes and, among them, the chimpanzees and bonobos, who are more closely related to us than they are to the gorillas, and much more than they are to the orangutans. The hominid lineage diverged from the chimpanzee lineage 6–8 million years ago (mya) and evolved exclusively in the African continent until the emergence of *Homo erectus*, somewhat before 1.8 mya. The first known hominids are *Sahelanthropus tchadensis* (dated 6–7 mya), *Orrorin tugenensis* (dated 5.8–6.1 mya),

141

and *Ardipithecus ramidus* (dated 5.2–5.8 mya). These hominids were, for the most part, bipedal when on the ground, but retained tree-climbing abilities and practices. It is not certain that they all are in the direct line of descent to modern humans, *Homo sapiens. Australopithecus anamensis*, dated 3.9–4.2 mya, was habitually bipedal and has been placed in the line of descent to *Australopithecus afarensis, Homo habilis, H. erectus*, and *H. sapiens*. Other hominids, not in the direct line of descent to modern humans, are *Australopithecus africanus, Paranthropus aethiopicus, P. boisei*, and *P. robustus*, who lived in Africa at various times between 3 and 1 mya, a period when three or four hominid species lived contemporaneously in the African continent (see Cela Conde and Ayala 2001, for a recent extensive review of hominid evolution).

The first intercontinental wanderer among our ancestors is *H. erectus*. Shortly after its emergence in tropical or subtropical eastern Africa, *H. erectus* dispersed to other continents of the Old World. Fossil remains of *H. erectus* are known from Africa, Indonesia (Java), China, the Middle East, and Europe. *H. erectus* fossils from Java have been dated 1.81 ± 0.04 and 1.66 ± 0.04 mya, and from Georgia between 1.6 and 1.8 mya. Anatomically distinctive *H. erectus* fossils have been found in Spain, deposited before 780,000 years ago, the oldest in western Europe.

Fossil remains of Neanderthal hominids (*Homo neanderthalensis*), with brains as large as those of *H. sapiens*, appeared in Europe around 200 thousand years ago (kya) and persisted until 40 kya. The Neanderthals were thought to be ancestral to anatomically modern humans, but now we know that modern humans appeared at least 100 kya, much before the disappearance of the Neanderthals. Moreover, in caves in the Middle East, fossils of modern humans have been found dated nearly 100 kya, as well as Neanderthals dated at 60 and 70 kya, followed again by modern humans dated at 40 kya. It is unclear whether the two forms repeatedly replaced one another by migration from other regions or whether they coexisted in the same areas. Recent genetic evidence indicates that interbreeding between *sapiens* and *neanderthalensis* never occurred.

The origin of anatomically modern humans is controversial. Some anthropologists argue that the transition from *H. erectus* to archaic *H. sapiens* and later to anatomically modern humans occurred consonantly in various parts of the Old World. Proponents of this "multiregional model" call attention to fossil regional continuity in the transition from *H. erectus* to archaic and then modern *H. sapiens*. They postulate that genetic exchange occurred from time to time between populations, so that the species evolved as a single gene pool, even though geographic differentiation occurred and persisted, just as geographically differentiated populations exist in other animal species. This explanation depends on the postulate of persistent migrations and interbreeding between populations from different continents, of which no direct evidence exists. Moreover, it is difficult to conciliate the multiregional model with fossil evidence of the contemporary existence of different species (*H. erectus* and *H. sapiens*) or forms (archaic and modern *H. sapiens*) in China, Indonesia, and other regions.

Other scientists argue instead that modern humans first arose in Africa (or in the Middle East) between 150 kya and 100 kya and from there spread throughout the world, replacing elsewhere the preexisting populations of *H. erectus* or archaic *H. sapiens*.

Genetic and molecular evidence show greater difference between African and non-African populations than between other human groups. This consistent differentiation endorses the hypothesis that the origin of anatomically modern humans was in Africa, whence modern humans expanded to the rest of the world starting about 100 kya. It is not possible, however, to exclude completely a partial participation of archaic *H. sapiens* from the Old World in the origin of modern humans. Two articles published in January 2001 evince the persistence in modern humans of older anatomical traits in populations of central Europe and of genetic traces (in the so-called mitochondrial DNA) from Australia (Wolpoff et al. 2001; Adcock et al. 2001). A recent genetic analysis supports the occurrence of at least two, not one, major migrations out of Africa, well after the original range expansion of *H. erectus* (Templeton 2002).

Much remains unknown about our evolutionary origins. The picture I have sketched is likely to be modified, as new discoveries are made. The discovery, announced in 2004, of a very small descendant of *H. erectus* in the Indonesian island of Flores is a case in point (Brown et al. 2004; Morwood et al. 2004). This *H. floresiensis* was only about one meter tall and had a very small cranial capacity, comparable in size with that of *Australopithecus*, yet it lived between 38 and 18 kya (and, probably, until 12 kya), well after individuals of our species, *H. sapiens*, were living in other parts of Australasia.

HUMANKIND'S DISTINCTIVE TRAITS

Erect posture and large brain are the two most conspicuous human anatomical traits. We are the only vertebrate species with a bipedal gait and erect posture; birds are bipedal, but their backbone stands horizontal rather than vertical (penguins are a minor exception). Brain size is generally proportional to body size; relative to body mass, humans have the largest (and most complex) brain. The chimpanzee's brain weighs less than a pound; a gorilla's slightly more. The human male adult brain has a volume of 1,400 cubic centimeters (cc), about three pounds in weight.

Until recently, evolutionists had raised the question whether bipedal gait or large brain came first, or whether they evolved consonantly. The issue is now resolved. Our *Australopithecus* ancestors had, since 4 mya, a bipedal gait but a small brain, about 450 cc, a pound in weight. Brain size starts to increase notably with our *Homo habilis* ancestors, about 2.5 mya, who had a brain about 650 cc and also were prolific toolmakers (hence the name *habilis*). Between 1 and 2 million years afterward, there lived *Homo erectus*, with adult brains about 1,200 cc. Our species, *Homo sapiens*, has a brain about three times as large as that of *Australopithecus*, 1,300–1,400 cc, or some three pounds of gray matter. Our brain is not only much larger than that of chimpanzees or gorillas, but also much more complex. The cerebral cortex, where the higher cognitive functions are processed, is in humans disproportionally much greater than the rest of the brain when compared with that in apes.

Erect posture and large brain are not the only anatomical traits that distinguish us from nonhuman primates, even if they may be the most obvious. A list of our most distinctive anatomical features includes the following:

- Erect posture and bipedal gait (which entail changes of the backbone, hipbone, and feet)
- Opposing thumbs and arm and hand changes (making possible precise manipulation)
- Large brain
- Reduction of jaws and remodeling of face
- Changes in skin and skin glands
- Reduction in body hair
- Cryptic ovulation (and extended female sexual receptivity)
- Slow development
- Modification of vocal tract and larynx
- Reorganization of the brain

Humans are notably different from other animals not only in anatomy but also, and no less importantly, in their behavior, both as individuals and socially. A list of distinctive human behavioral traits includes the following:

- Subtle expression of emotions
- Intelligence: abstract thinking, categorizing, and reasoning

- Symbolic (creative) language
- Self-awareness and death awareness
- Toolmaking and technology
- Science, literature, and art
- Ethics and religion
- Social organization and cooperation (division of labor)
- Legal codes and political institutions

Humans live in groups that are socially organized, and so do other primates. But primate societies do not approach the complexity of human social organization. A distinctive human social trait is culture, which may be understood as the set of nonstrictly biological human activities and creations. Culture includes social and political institutions, ways of doing things, religious and ethical traditions, language, common sense and scientific knowledge, art and literature, technology, and in general all the creations of the human mind. The advent of culture has brought with it cultural evolution, a superorganic mode of evolution superimposed on the organic mode, which has, in the last few millennia, become the dominant mode of human evolution. Cultural evolution has come about because of cultural change and inheritance, a distinctively human mode of achieving adaptation to the environment and transmitting it through the generations.

BIOLOGICAL EVOLUTION VERSUS CULTURAL EVOLUTION

There are in humankind two kinds of heredity – the biological and the cultural, which may also be called organic and superorganic, or endosomatic and exosomatic systems of heredity. Biological inheritance in humans is very much like that in any other sexually reproducing organism; it is based on the transmission of genetic information encoded in DNA from one generation to the next by means of the sex cells.

Cultural inheritance, in contrast, is based on transmission of information by a teaching-learning process, which is in principle independent of biological parentage. Culture is transmitted by instruction and learning; by example and imitation; through books, newspapers and radio, television and motion pictures; through works of art; and by any other means of communication. Culture is acquired by every person from parents, relatives, and neighbors and from the whole human environment (Dobzhansky 1962; Ehrlich 2000).

Cultural inheritance makes possible for humans what no other organism can accomplish – the cumulative transmission of experience from generation to generation. Animals can learn from experience, but they do not transmit their experiences, their "discoveries" (at least not to any large extent) to the following generations. Animals have individual memory, but they do not have a "social memory." Humans, on the other hand, have developed a culture because they can transmit cumulatively their experiences from generation to generation.

Cultural inheritance makes possible cultural evolution, that is, the evolution of knowledge, social structures, ethics, and all other components that make up human culture. Cultural inheritance makes possible a new mode of adaptation to the environment that is not available to nonhuman organisms – adaptation by means of culture. Organisms in general adapt to the environment by means of natural selection, by changing over generations their genetic constitution to suit the demands of the environment. But humans, and humans alone, can also adapt by changing the environment to suit the needs of their genes. (Animals build nests and modify their environment also in other ways, but the manipulation of the environment by any nonhuman species is trivial compared with humankind's.)

For the past few millennia, humans have been adapting the environments to their genes more often than their genes to the environments. In order to extend its geographical habitat, or to survive in a changing environment, a population of organisms must become adapted, through slow accumulation of genetic variants sorted out by natural selection, to the new climatic conditions, different sources of food, different competitors, and so on. The discovery of fire and the use of shelter and clothing allowed humans to spread from the warm tropical and subtropical regions of the Old World to the whole Earth, except for the frozen wastes of Antarctica, without the anatomical development of fur or hair. Humans did not wait for genetic mutants promoting wing development; they have conquered the air in a somewhat more efficient and versatile way by building flying machines. People travel the rivers and the seas without gills or fins. The exploration of outer space has started without waiting for mutations providing humans with the ability to breathe under low oxygen pressures or to function in the absence of gravity; astronauts carry their own oxygen and wear specially equipped pressure suits. From their obscure beginnings in Africa, humans have become the most widespread and abundant species of mammal on earth. It was the appearance of culture as a superorganic form of adaptation that made humankind the most successful animal species.

Cultural adaptation is more effective than biological adaptation because it is more rapid and because it can be directed. A favorable genetic mutation newly arisen in an individual can be transmitted to a sizable part of the human species only through innumerable generations. However, a new scientific discovery or technical achievement can be transmitted to the whole of humankind, potentially at least, in less than one generation. Moreover, whenever a need arises, humans can directly pursue the appropriate cultural changes to meet the challenge. On the contrary, biological adaptation depends on the accidental availability of a favorable mutation, or of a combination of several mutations, at the time and place where the need arises.

High intelligence, symbolic language, religion, and ethics are some of the behavioral traits that distinguish us from other animals. The account of human origins that I have sketched here implies a continuity in the evolutionary process that goes from our nonhuman ancestors of 8 mya, through primitive hominids, to modern humans. A scientific explanation of that evolutionary sequence must account for the emergence of human anatomical and behavioral traits in terms of natural selection, together with other distinctive biological causes and processes. One explanatory strategy is to focus on a particular human feature and seek to identify the conditions under which this feature may have been favored by natural selection. Such a strategy may lead to erroneous conclusions as a consequence of the fallacy of selective attention: some traits may have come about not because they are themselves adaptive, but rather because they are associated with traits that are favored by natural selection.

Genes that become changed by natural selection owing to their effects on a certain trait may result in the modification of other traits as well, even if these additional changes are neutral to natural selection. The changes of these other traits are epigenetic (or "pleiotropic," in even more esoteric genetic jargon) consequences of the changes directly promoted by natural selection. The cascade of consequences may be, particularly in the case of humans, very long and far from obvious in some cases. Literature, art, science, technology, ethics, and religion are among the behavioral features that may have come about not because they were adaptively favored in human evolution but because they are expressions of the high intellectual abilities present in modern humans: what may have been favored by natural selection (its "target") was an increase in intellectual ability rather than each one of those particular activities.

ETHICAL BEHAVIOR VERSUS ETHICAL NORMS

Ethics and ethical behavior may serve as a model case of how we may seek the evolutionary explanation of a distinctively human trait. The objective is to ascertain whether an account can be advanced of ethical behavior as an outcome of biological evolution and, if such is the case, whether ethical behavior was directly promoted by natural selection, or has rather come about as an epigenetic manifestation of some other trait that was the target of natural selection.

The question whether ethical behavior is biologically determined may refer either to the *capacity* for ethics (i.e., the proclivity to judge human actions as either right or wrong) and which I refer to as "ethical behavior"; or the moral *norms* or moral codes accepted by human beings for guiding their actions. A similar distinction can be made with respect to language. The issue whether the capacity for symbolic language is determined by our biological nature is different from the question of whether the particular language we speak (English, Spanish, or Japanese) is biologically necessary.

The first question posed asks whether the biological nature of *Homo sapiens* is such that humans are necessarily inclined to make moral judgments and to accept ethical values, to identify certain actions as either right or wrong. Affirmative answers to this first question do not necessarily determine what the answer to the second question should be. Independently of whether or not humans are necessarily ethical, it remains to be determined whether particular moral prescriptions are in fact determined by our biological nature, or whether they are chosen by society, or by individuals. Even if we were to conclude that people cannot avoid having moral standards of conduct, it might be that the choice of the particular standards used for judgment would be arbitrary or that it depended on some other, nonbiological criteria. The need for having moral values does not necessarily tell us what these moral values should be, just as the capacity for language does not determine which language we shall speak.

The thesis I propose is that humans are ethical beings by their biological nature. Humans evaluate their behavior as either right or wrong, moral or immoral, as a consequence of their eminent intellectual capacities, which include self-awareness and abstract thinking. These intellectual capacities are products of the evolutionary process, but they are distinctively human. Thus, I maintain that ethical behavior is not causally related to the social behavior of animals, including kin and reciprocal "altruism" (Ayala 1987, 1995).

A second thesis is that the moral norms according to which we evaluate particular actions as morally either good or bad (as well as the grounds that may be used to justify the moral norms) are products of cultural evolution, not of biological evolution. The norms of morality belong, in this respect, to the same category of phenomena as the languages spoken by different peoples, their political and religious institutions, and the arts, sciences, and technology. The moral codes are in some respects isomorphic with the biological predispositions of the human species, dispositions we share to some extent with other animals. But this isomorphism between ethical norms and biological tendencies is not necessary or universal: it does not apply to all ethical norms in a given society, much less in all human societies.

This second thesis contradicts the proposal of many distinguished evolutionists who, since Darwin's time, have argued that the norms of morality are derived from biological evolution. It also contradicts the sociobiologists, who have recently developed a subtle version of that proposal. The sociobiologists' argument is that human ethical norms are sociocultural correlates of behaviors fostered by biological evolution. I argue that such proposals are misguided and do not escape the naturalistic fallacy. It is true that both natural selection and moral norms sometimes target the same behavior; that is, the two are consistent. But this consistency between the behaviors promoted by natural selection and those sanctioned by moral norms exists only with respect to the consequences of the behaviors; the underlying causations are completely disparate.

Moral codes, like any other dimensions of cultural systems, depend on the existence of human biological nature and must be consistent with it in the sense that they could not counteract it without promoting their own demise. Moreover, the acceptance and persistence of moral norms is facilitated whenever they are consistent with biologically conditioned human behaviors. But the moral norms are independent of such behaviors in the sense that some norms may not favor, and may hinder, the survival and reproduction of the individual and its genes, which are the targets of biological evolution. Discrepancies between accepted moral rules and biological survival are, however, necessarily limited in scope or would otherwise lead to the extinction of the groups accepting such discrepant rules.

BIOLOGICAL ROOTS OF ETHICAL BEHAVIOR

The question whether ethical behavior is determined by our biological nature must be answered in the affirmative. By "ethical behavior" I mean here to refer to the judging of human actions as either good or bad, which is not the same as "good behavior" (i.e., doing what is perceived as good instead of what is perceived as evil). Humans exhibit ethical behavior by nature because their biological constitution determines the presence of the three necessary conditions for ethical behavior. These conditions are the ability to anticipate the consequences of one's own actions; the ability to make value judgments; and the ability to choose between alternative courses of action. I briefly examine each of these abilities and show that they are consequences of the eminent intellectual capacity of human beings.

The ability to anticipate the consequences of one's own actions is the most fundamental of the three conditions required for ethical behavior. Only if I can anticipate that pulling the trigger will shoot the bullet, which in turn will strike and kill my enemy, can the action of pulling the trigger be evaluated as nefarious. Pulling a trigger is not in itself a moral act; it becomes so by virtue of its relevant consequences. My action has an ethical dimension only if I do anticipate these consequences.

The ability to anticipate the consequences of one's actions is closely related to the ability to establish the connection between means and ends – that is, of seeing a means precisely as means, as something that serves a particular end or purpose. This ability to establish the connection between means and their ends requires the ability to anticipate the future and to form mental images of realities not present or not yet in existence.

The ability to establish the connection between means and ends happens to be the fundamental intellectual capacity that has made possible the development of human culture and technology. A reasonable evolutionary hypothesis to account for this capacity proposes that its roots may be found in the evolution of bipedal gait, which transformed the anterior limbs of our ancestors from organs of locomotion into organs of manipulation. The hands thereby gradually became organs adept for the construction and use of objects for hunting and other activities that improved survival and reproduction.

The construction of tools, however, depends not only on manual dexterity but on perceiving them precisely as tools, as objects that help to perform certain actions, that is, as means that serve certain ends or purposes: a knife for cutting, an arrow for hunting, an animal skin for protecting the body from the cold. The hypothesis I am propounding is that natural selection promoted the intellectual capacity of our biped ancestors because increased intelligence facilitated the perception of tools as tools, and therefore their construction and use, with the ensuing amelioration of biological survival and reproduction.

The development of the intellectual abilities of our ancestors took place over 2 million years or longer, gradually increasing the ability to connect means with their ends and, hence, the possibility of making ever more complex tools serving remote purposes. The ability to anticipate the future, essential for ethical behavior, is therefore closely associated with the development of the ability to construct tools, an ability that has produced the advanced technologies of modern societies and that is largely responsible for the success of humankind as a biological species.

The second condition for the existence of ethical behavior is the ability to make value judgments, to perceive certain objects or deeds as more desirable than others. Only if I can see the death of my enemy as preferable to his or her survival (or vice versa) can the action leading to his or her demise be thought of in moral terms. If the alternative consequences of an action are neutral with respect to value, the action does not belong within the scope of ethical behavior. The ability to make value judgments depends on the capacity for abstraction, that is, on the capacity to perceive actions or objects as members of general classes. This makes it possible to compare objects or actions with one another and to perceive some as more desirable than others. The capacity for abstraction, necessary to perceive individual objects or actions as members of general classes, requires an advanced intelligence such as it exists only in humans. Thus, I see the ability to make value judgments primarily as an implicit consequence of the enhanced intelligence favored by natural selection in human evolution. Nevertheless, valuing certain objects or actions and choosing them over their alternatives can be of biological consequence; doing this in terms of general categories can be beneficial in practice.

Value judgments indicate preference for what is perceived as good and rejection of what is perceived as bad; good and bad may refer to economic, aesthetic, or all sorts of other kinds of values. Moral judgments concern the values of right and wrong in human conduct. Moral judgments are a particular class of value judgments; namely those where preference is not dictated by one's own interest or profit but by regard for others, which may cause benefits to particular individuals (altruism) or take into consideration the interests of a social group to which one belongs.

Evolutionists have demonstrated that "group selection" is not an "evolutionary stable strategy." Group selection refers to selection that benefits the group at the expense of the (inclusive) fitness of the individual. Suppose that there is a group with a genetic trait that benefits the group to the extent that the group is very successful and expands in numbers at the expense of other groups and to the benefit and multiplication of the individuals in the group and their genetic makeups. Suppose now that a mutation arises in an individual that makes it behave selfishly. Individuals carrying this mutation will benefit from the altruistic behavior of the others and will not incur the costs of the others' altruistic behavior. Consequently the selfish individuals will have higher fitness than the altruists and the selfish mutation will increase in frequency until it eliminates from the group the altruistic gene.

Humans, however, can perceive the benefits of altruistic behavior for the group (and through the group to themselves) and choose to behave altruistically. The altruistic behavior may be enforced by political authority by imposing a penalty (if you commit adultery, or if you steal, you'll be stoned to death, or jailed, or otherwise punished) or promoted through religious authority or belief, like the Christian commandments against adultery and theft. Thus, morality makes it possible for true altruism to be an evolutionary stable strategy. But this depends on humans' exalted intelligence and the presence of the three conditions for moral behavior.

The third condition necessary for ethical behavior is the ability to choose between alternative courses of action. Pulling the trigger can be a moral action only if I have the option not to pull it. A necessary action beyond our control is not a moral action: the circulation of the blood or the digestion of food are not moral actions.

Whether there is free will has been much discussed by philosophers, and this is not the appropriate place to review the arguments. I only advance two considerations based on commonsense experience. One is our profound personal conviction that the possibility of choosing between alternatives is genuine rather than only apparent. The second consideration is that when we confront a given situation that requires action on our part, we are able mentally to explore alternative courses of action, thereby extending the field within which we can exercise our free will. In any case, if there were no free will, there would be no ethical behavior; morality would only be an illusion. The point I wish to make here is, however, that free will is dependent on the existence of a well-developed intelligence, which makes it possible to explore alternative courses of action and to choose one or another in view of the anticipated consequences.

In summary, ethical behavior is an attribute of the biological makeup of humans and is, in that sense, a product of biological evolution. But I see no evidence that ethical behavior developed because it was adaptive in itself. I find it hard to see how *evaluating* certain actions as either good or evil (not just choosing some actions rather than others, or evaluating them with respect to their practical consequences) would promote the reproductive fitness of the evaluators. Nor do I see how there might be some form of "incipient" ethical behavior that would then be further promoted by natural selection.

It seems rather that the likely target of natural selection was the development of advanced intellectual capacities. This development was favored by natural selection because the construction and use of tools improved the strategic position of our biped ancestors. Once bipedalism evolved and tool-using and toolmaking became possible, those individuals more effective in these functions had a greater probability of biological success. The biological advantage provided by the design and use of tools persisted long enough so that intellectual abilities continued to increase, eventually yielding the eminent development of intelligence that is characteristic of *Homo sapiens.*

ETHICAL NORMS: BEYOND BIOLOGY

Since the publication of Darwin's theory of evolution by natural selection, philosophers as well as biologists have attempted to find in the evolutionary process the justification for moral norms. The common ground to such proposals is that evolution is a natural process that achieves goals that are desirable and thereby morally good; indeed it has produced humans (Ayala 1987). Proponents of these ideas claim that only the evolutionary goals can give moral value to human action: whether a human deed is morally right depends on whether it directly or indirectly promotes the evolutionary process and its natural objectives.

A different attempt to ground moral codes on the evolutionary process is that of the sociobiologists, particularly from E. O. Wilson (1975, 1978; see also Alexander 1987), who starts by proposing that "scientists and humanists should consider together the possibility that the time has come for ethics to be removed temporarily from the hands of the philosophers and biologicized" (1975, p. 562). The sociobiologists argue that the perception that morality exists is an epigenetic manifestation of our genes, which so manipulate humans as to make them believe that some behaviors are morally "good" so that people behave in ways that are good for their genes. Humans might not otherwise pursue these behaviors – altruism, for example – because their genetic benefit is not apparent (except to sociobiologists after the development of their discipline) (Ruse 1986a, b; Ruse and Wilson 1986).

Wilson writes: "Human behavior – like the deepest capacities for emotional response which drive and guide it – is the circuitous technique by which human genetic material has been and will be kept intact. *Morality has no other demonstrable ultimate function*" (E. O. Wilson 1978, p. 167; emphasis added). How is one to interpret this statement? It is possible that Wilson is simply giving the reason why ethical behavior exists at all, in the sense I have just stated; namely, our genes prompt us to accept what we call "morality," so that we act accordingly to the interests of our genes, interests that are not otherwise apparent to us.

It is possible, however, to read Wilson's statement as a justification of human moral codes: the function of these would be to preserve human genes. But this would entail the naturalistic fallacy and, worse yet, would seem to justify a morality that most of us detest. If the preservation of human genes (be those of the individual, the group, or the species) is the purpose that moral norms serve, Spencer's social Darwinism would seem right; racism or even genocide could be justified as morally correct if they were perceived as the means to preserve those genes thought to be good or desirable and to eliminate those thought to be bad or undesirable. There is no doubt in my mind that Wilson is not intending to justify racism or genocide, but this is one possible interpretation of his words.

I now turn to the sociobiologists' proposition that natural selection favors behaviors that are isomorphic with the behaviors sanctioned by the moral codes endorsed by most humans.

Evolutionists had for years struggled with finding an explanation for the apparently altruistic behavior of animals. When predators attack a herd of zebras, these zebras will attempt to protect the young in the herd, even if they are not their progeny, rather than fleeing. When a prairie dog sights a coyote, it will warn other members of the colony with an alarm call, even though by drawing attention to itself this increases its own risk. Examples of altruistic behaviors of this kind can be multiplied.

Altruism is defined in the dictionary I happen to have at hand (*Merriam Webster's Collegiate Dictionary*, 10th ed.) as "unselfish regard for, or devotion to the welfare of others." The dictionary gives a second definition: "behavior by an animal that is not beneficial to or may be harmful to itself but that benefits others of its species." To speak of animal altruism is not to claim that explicit feelings of devotion or regard are present in them, but rather that animals act for the welfare of others at their own risk just as humans are expected to do when behaving altruistically. The problem is precisely how to justify such behaviors in terms of natural selection. Assume, for illustration, that in a certain species there are two alternative forms of a gene ("alleles"), of which one but not the other promotes altruistic behavior. Individuals possessing the altruistic allele will risk their life for the benefit of others, whereas those possessing the nonaltruistic allele will benefit from altruistic behavior without risking themselves. Possessors of the altruistic allele will be more likely to die, and the allele will therefore be eliminated more often than the nonaltruistic allele. Eventually, after some generations, the altruistic allele will be completely replaced by the nonaltruistic one. But then how is it that altruistic behaviors are common in animals without the benefit of ethical motivation?

One major contribution of sociobiology to evolutionary theory is the notion of "inclusive fitness." In order to ascertain the consequences of natural selection, it is necessary to take into account a gene's effects not only on a particular individual but on all individuals possessing that gene. When considering altruistic behavior, one must take into account not only the risks for the altruistic individual, but also the benefits for other possessors of the same allele. Zebras live in herds where individuals are blood relatives. A gene prompting adults to protect the defenseless young would be favored by natural selection if the benefit (in terms of saved carriers of that gene) is greater than the cost (due to the increased risk of the protectors). An individual that lacks the altruistic gene and carries instead a nonaltruistic one, will not risk its life, but the nonaltruistic allele is partially eradicated with the death of each defenseless relative.

It follows from this line of reasoning that the more closely related the members of a herd or animal group are, the more altruistic behaviors should be present. This seems to be generally the case. We need not enter here into the details of the quantitative theory developed by sociobiologists in order to appreciate the significance of two examples. The most obvious is parental care. Parents feed and protect their young because each child has half the genes of each parent: the genes are protecting themselves, as it were, when they prompt a parent to care for its young.

A second example is more subtle: the social organization and behavior of certain animals like the honeybee. Worker bees toil building the hive and feeding and caring for the larvae even though they themselves are sterile and only the queen produces progeny. Assume that in some ancestral hive, a gene arises that prompts worker bees to behave as they now do. It would seem that such a gene would not be passed on to the following generation because such worker bees do not reproduce. But such inference is erroneous. Queen bees produce two kinds of eggs: some that remain unfertilized develop into males (which are therefore "haploid," i.e., carry only one set of genes); others that are fertilized (hence, are "diploid," carry two sets of genes) develop into worker bees and occasionally into a queen. W. D. Hamilton (1964) demonstrated that with such a reproductive system daughter queens and their worker sisters share in two-thirds of their genes, whereas daughter queens and their mother share in only one-half of their genes. Hence, the worker bee genes are more effectively propagated by workers caring for their sisters than if they would produce and care for their own daughters. Natural selection can thus explain the existence in social insects of sterile casts, which exhibit a most extreme form of apparently altruistic behavior by dedicating their life to care for the progeny of another individual (the queen). The theory predicts that the hive will tend to minimize the number of reproductive females, which is what happens in the honeybee, where all reproduction is performed by the one queen.

Sociobiologists point out that many of the moral norms commonly accepted in human societies sanction behaviors also promoted by natural selection (which promotion becomes apparent only when the inclusive fitness of genes is taken into account). Examples of such behaviors are the commandment to honor one's parents, the incest tabu, the greater blame attributed to the wife's than to the husband's adultery, the ban or restriction on divorce, and many others. The sociobiologists' argument is that human ethical norms are sociocultural correlates of behaviors fostered by biological evolution. Ethical norms protect such evolution-determined behaviors as well as being specified by them.

I believe, however, that the sociobiologists' argument is misguided and does not escape the naturalistic fallacy (Ayala 1987, 1995; see also Sober and Wilson 1998). Consider altruism as an example. Altruism in the biological sense (altruism$_b$) is defined in terms of the population genetic consequences of a certain behavior. Altruism$_b$ is explained by the fact that genes prompting such behavior are actually favored by natural selection (when inclusive fitness is taken into account), even though the fitness of the behaving individual is decreased. But altruism in the moral sense (altruism$_m$) is explained in terms of motivations: a person chooses to risk his own life (or incur some kind of "cost") for the benefit of somebody else. The isomorphism between altruism$_b$ and altruism$_m$ is only apparent: an individual's chances are improved by the behavior of another individual who incurs a risk or cost. The underlying causations are completely disparate: the ensuing genetic benefits in altruism$_b$; and regard for others in altruism$_m$. (Sociobiologists, however, might say that our perception that our altruistic behavior is motivated by regard for others is itself caused by our genes that seek that way to accomplish their own purposes. As Ruse (1986b) puts it: "[Sociobiologists] argue that moral (literal) altruism might be one way in which biological (metaphorical) 'altruism' could be achieved. . . . Literal, moral altruism is a major way in which advantageous biological cooperation is achieved. . . . In order to achieve [biological] 'altruism' we are altruistic." This is, of course, a claim of biological determinism in the extreme and ultimately entails the denial of true free will.)

One additional observation worth notice is that some norms of morality are consistent with behaviors prompted by natural selection, but other norms are not so. The commandment of charity, "Love thy neighbor as thyself," often runs contrary to the inclusive fitness of the genes, even though it promotes social cooperation and peace of mind. If the yardstick of morality were the multiplication of genes, the supreme moral imperative would be to beget the largest possible number of children and (with lesser dedication) to encourage our close relatives to do the same. But to impregnate the most women possible is not, in the view of most people, the highest moral duty of a man.

Biology to Ethics

As I stated earlier, my view is that we make moral judgments as a consequence of our eminent intellectual abilities, not as an innate way for achieving
biological gain, and that the codes of morality by which humans guide their
actions have been formulated as a consequence of social traditions and/or
by religious or political authority (D. S. Wilson 2002). The codes of morality that prevail in human populations are largely consistent with the genetic
interests of individuals because otherwise the codes would not have historically survived. But the codes also incorporate norms that benefit the tribe
or community rather than the individual, such as the commandments against
adultery or theft.

I summarize my views by returning to the analogy with human languages.
Our biological nature determines the sounds that we can or cannot utter and
also constrains human language in other ways. But a language's syntax and
vocabulary are not determined by our biological nature (otherwise, there
could not be a multitude of tongues), but are products of human culture.
Likewise, moral norms are determined not by biological processes but by
cultural traditions and principles including religious beliefs that are products
of human history.

REFERENCES

Adcock, G. J., Dennis, E. S., Easteal, S., Huttley, G. A., Jermiin, L. S., and Peacock,
    W. J. 2001. Mitochondrial DNA Sequences in Ancient Australians: Implications for
    Modern Human Origins. *Proc. Natl. Acad. Sci. USA* 98: 537–542.
Alexander, R. D. 1987. *The Biology of Moral Systems*. Hawthorne, N.Y.: Aldine.
Ayala, F. J. 1987. The Biological Roots of Morality. *Biology and Philosophy* 2: 235–252.
Ayala, F. J. 1995. The Difference of Being Human: Ethical Behavior as an Evolutionary
    Byproduct. In *Biology, Ethics and the Origin of Life*, ed. H. Rolston III, 113–135.
    Boston and London: Jones and Bartlett.
Brown, P., Sutikna, T., Morwood, M. J., Soejono, R. P., Jatmiko, Wayhu, Saptomo, E.,
    and Rokus Awe Due . 2004. A New Small-Bodied Hominin from the Late Pleistocene
    of Flores, Indonesia. *Nature* 431: 1055–1061.
Cela Conde, C. J., and Ayala, F. J. 2001. *Senderos de la Evolución Humana*. Madrid:
    Alianza Editorial.
Dobzhansky, T. 1962. *Mankind Evolving: The Evolution of the Human Species*. New
    Haven: Yale University Press.
Ehrlich, P. R. 2000. *Human Natures: Genes, Cultures, and the Human Prospect*. Wash-
    ington, D.C., and Covelo, Calif.: Island Press/Shearwater Books.
Hamilton, W. D. 1964. The Genetic Evolution of Social Behavior. *J. Theoretical Biology*
    7: 1–52.
Morwood, M. J., Soejono, R. P., Roberts, R. G., Sutikna, T., Turney, C. S. M., Westaway,
    K. E., Rink, W. R., Zhao, J.-X., Van Den Bergh, G. D., Rokus Awe Due, Hobbs,

D. R., Moore, M. W., Bird, M. I., and Fifield, L. K. 2004. Archaeology and Age of a New Hominin from Flores in Eastern Indonesia. *Nature* 431: 1087–1091.

Ruse, M. 1986a. *Taking Darwin Seriously: A Naturalistic Approach to Philosophy.* Oxford: Basil Blackwell.

Ruse, M. 1986b. Evolutionary Ethics: A Phoenix Arisen. *Zygon* 21: 95–112.

Ruse, M., and Wilson, E. O. 1986. Moral Philosophy as Applied Science. *Journal of the Royal Institute of Philosophy* 61: 173–192.

Sober, E., and Wilson, D. S. 1998. *Unto Others: The Evolution and Psychology of Unselfish Behavior.* Cambridge, Mass.: Harvard University Press.

Templeton, A. R. 2002. Out of Africa Again and Again. *Nature* 416: 45–51.

Wilson, D. S. 2002. *Darwin's Cathedral: Evolution, Religion, and the Nature of Society.* Chicago: University of Chicago Press.

Wilson, E. O. 1975. *Sociobiology: The New Synthesis.* Cambridge, Mass.: Harvard University Press.

Wilson, E. O. 1978. *On Human Nature.* Cambridge, Mass.: Harvard University Press.

Wolpoff, M. H., Hawks, J., Frayer, D. W., and Hunley, K. 2001. Modern Human Ancestry at the Peripheries: A Test of the Replacement Theory. *Science* 291: 293–297.

9

Between Fragile Altruism and Morality

Evolution and the Emergence of Normative Guidance

PHILIP KITCHER

BIOLOGICAL ALTRUISM

Philosophical discussions of the relationship between biology and morality often gravitate to one of two positions. At the Hobbesian pole are views claiming that human nature is fundamentally selfish and that morality is a system that restrains and runs contrary to our most basic impulses. The Humean pole, by contrast, develops the idea that human beings have a natural disposition to fellow feeling, one that our moral sentiments are able to extend and refine. I hope to explain, and partially defend, a position that combines elements of both perspectives.

Both Hobbesians and Humeans take a stand on the question of altruism, and it is natural to start from our current biological understanding of the possibility (or possibilities) of altruism. For evolutionary biologists, of course, altruistic behavior is understood as behavior that increases the reproductive success of another organism at reproductive cost to the beneficiary.[1] Throughout the twentieth century, biology faced the theoretical problem of how to reconcile the possibility of altruistic behavior with Darwinian natural selection. I am not going to dwell on the details of the history. It is enough to note that, by the end of the century, the problem had been largely solved. Thanks to theories of kin selection and reciprocal altruism, to evolutionary game theory, and to a refined version of group selection, there are ample devices for showing the *general* possibility that altruistic behavior can originate and be maintained under natural selection.[2]

---

[1] This is a simplified formulation, but it will do for our purposes. A more accurate account would speak of expected reproductive benefits and costs.

[2] See Hamilton 1971, Trivers 1971, Axelrod 1984, Maynard-Smith 1982, and Sober and Wilson 1998. I should note here that, while I agree with Sober and Wilson's claim that an account of group selection can be developed to avoid the traditional difficulties of that conception, it seems to

That does not mean, however, that questions about the evolution of human altruism have been adequately resolved. There are two difficulties. First, the-oretical models for explaining the possibility of biological altruism can be applied to our own lineage only if there are ways of linking the environments experienced by our ancestors to the conditions postulated in those models; the well-known Axelrod-Hamilton dynamics would only account for altruism in hominids if there were some period in our past at which primates, great apes, or early humans were forced into pairwise interactions of the form of an indefinitely repeated prisoners' dilemma. Second, when we confront the issue that divides Hobbesians from Humeans, the notion of altruism that concerns us is not that of the biologists; the important concept is psychological, having little intrinsic connection with reproductive success and everything to do with the intentions of the agent.

Some years ago, I attempted to address the first difficulty by showing how a variant of the Axelrod-Hamilton model would yield an apparently more realistic account of hominid altruism (in the biological sense). Instead of envisaging a mechanism that paired organisms for repeated interactions (as if a gigantic hand came down on the savannah and forced two of our ancestors together for PD play!), I supposed that populations of organisms can encounter many situations in which it is possible to pursue a goal by themselves or to team up with others. An *optional game* is defined by the possibilities of both choosing to interact at all and choosing one's partner. As mathematically minded biologists and economists have seen (although philosophers have sometimes missed the point), the dynamics of optional games is different from that of compulsory games, and, indeed, it is possible to show that altruism emerges more readily and is present at higher frequencies in scenarios governed by the framework of optional games.[3] It is not hard to show that some cooperative primate behavior – for example, social grooming – can be modeled as an optional game, and this raised the hope that one could find many motors for human altruism in our prehistory.

That hope was dashed by a closer look at the social systems of our evolu-tionary relatives and by the available data on the ecological parameters that pertain to likely applications of the optional games framework to hominid

---

me that the attempt to understand all the ways of approaching the theoretical problem of altruism in terms of group selection can only succeed if the notion of a group is stretched so broadly that the thesis loses content.

[3] My approach was offered in Kitcher 1993 and in Batali and Kitcher 1995. John Batali and I showed how altruism would be more prevalent in the optional variety. Sober and Wilson (1998) seem quite unaware of the differences.

evolution. On the latter score, it is not hard to show that patterns of cooperative hunting among chimpanzees conform very poorly to the deliverances of the models: hunting does not appear to be nutritionally very important, and the distributions of rewards from group hunting seem to violate the theoretical expectations (nonparticipants frequently obtain rewards, while those who initiate a cooperative venture sometimes get nothing and yet are willing to team up again with those who appear to have "defected" on them). Even the seeming success of social grooming turns out to be a source of puzzles, for the hygienic benefits actually received are disproportionately small in relation to the amount of time invested. The overwhelmingly obvious fact that results from viewing primate social behavior through the lens of optional games is that the organisms involved are only playing such games in an evolutionarily sensible way if one supposes that there is a background social structure that adjusts the costs and benefits.

That conclusion is reinforced when we consider the variety of social structures found among the great apes, ranging from the mostly solitary behavior of orangutans through the family groups of the gibbons,[4] to the small bands of gorillas, typically dominated by a single male. Before one even applies the machinery of optional games – or anything similar – it is important to understand the emergence of the population within which the opportunities for cooperation are to arise. One cannot simply think of a population of organisms *who tolerate one another's presence* as given. The fundamental problem, I suggest, is how hominids became sufficiently at ease with one another to form a potentially cooperative population in the first place – what is the explanation of the difference between chimpanzees, bonobos, and hominids, on the one hand, and orangutans, on the other? (There are excellent reasons for thinking that our primitive social state was much like that of living chimpanzee groups.)[5]

The question just posed can be addressed by generalizing an approach offered by Richard Wrangham.[6] Imagine an environment in which resources are distributed, and suppose that organisms can only visit a small number

---

[4] Strictly speaking, of course, gibbons do not count as great apes; I include them here because the larger group might make a more natural unit for social comparisons.

[5] For example, the data available from hominid encampments suggest a group size with age and sex distributions much like that found in contemporary chimpanzees. A detailed account of the evidence for the chimp-human comparison is available in chapter 2 of Maryanski and Turner 1992. (I am indebted to Alex Rosenberg for mentioning this book to me; in many respects it develops a picture similar to the one I offer here.)

[6] See, especially, Wrangham 1987.

of such resources in each period of their lives. When organisms meet in the vicinity of a resource, the competition is decided by a variant of the hawk-dove game in which agents are able to detect one another's strength (the hawk-dove-assessor game). Organisms have different strengths, and we may suppose that every organism spends part of its life as relatively weak. It is not hard to show that, under some conditions, an initial solitary state would be preserved. For a wide variety of parameter settings, however, mathematical analysis and computer simulations both reveal that coalitions are likely to form, that there will be an escalation of coalition size, that the process will eventually terminate with a large coalition controlling a territory, and that within this large coalition there will be a nested subcoalitional structure, with relatively stable smaller units. In short, from an initial solitary state, a chimpanzee-hominid social structure can emerge.

The coalition game is mathematically intractable in that, to the best of my knowledge, there is no way to specify an optimal strategy; further, small perturbations of parameter values can shift a strategy from being a rather good one to being disastrous. The game is beyond the calculational abilities of economists and mathematicians – and, I will assume, of our primate ancestors whose information about their environment was far from perfect. Yet that is the game our ancestors had to play, or so I believe. The outcome of their efforts was a coalitional structure that affected the payoffs for the cooperative ventures that subsequently arose, making it a matter of extreme evolutionary importance to maintain, or improve, one's position in the available cluster of coalitions.

I have sketched (and *only* sketched) an account of the evolution of altruistic behavior (in the biologist's sense of "altruism") that might be applicable to organisms like us. If that sketch can be completed (as I think it can) we have tied up one of the loose ends left dangling earlier. Undertaking the project I announced at the beginning requires that we also come to terms with the different, and more difficult, notion of psychological altruism.

PSYCHOLOGICAL ALTRUISM

The altruism that matters to morality (and to moral philosophy) consists in a tendency to respond to the perceived needs and wants of others. To a first approximation, an altruist is someone who appreciates aspects of the lives of another person (or other people) and whose psychological states adjust in ways that dispose her to give aid. No action, no psychological state, is intrinsically altruistic; rather a person's altruism is reflected in a pattern of relationships among states and among states and actions.[7] In a full account, it would be important to consider a variety of states, looking at altruistic emotions and intentions; here I concentrate on altruistic desires.

Start with a paradigm. A person enters a room and finds within it a divisible good, one that she would like to consume in its entirety. If there was nobody else around, then she would consume it all. In some circumstances, however, her wants are changed; when certain people are also around (in the room or in the vicinity), she prefers a situation in which the good is divided to one in which she consumes it all. Her new preference comes about as a result of her recognizing the presence of those people and their wish to have a part of the good. She has one desire in one circumstance (herself alone) and a different desire in a closely related circumstance (just like the first except for the presence of another, or others), and the difference is explained by her perception of the desires of the other(s).

Altruists are not the only people who behave in this way. Such behavior is consistent with the account I gave that the focal character was moved by the hope that sharing would be reciprocated on some future occasion, or by the thought of impressing a third party in ways that would bring benefits down the road (or by any of a host of similar considerations). A genuine psychological altruist must meet a *noncalculational* requirement. That is, the presence of her new desire must not be explicable by supposing that she sees the outcome now desired as a means of promoting the satisfaction of other unmodified preferences.

So we can formulate a general account of psychologically altruistic desire (or preference):

A has weakly altruistic preferences toward *B* in circumstance *C* just in case:

1. There's a circumstance C^* like *C* except for the absence of any impact on *B*'s preferred outcomes, such that *A*'s preferences in *C* conform more closely to those she takes *B* to have than do *A*'s preferences in C^*.
2. The difference between the two preference structures is explained by *A*'s perception of *B*'s preferences.
3. Calculation of future benefits for *A* in terms of the preferences structure operative in C^* plays no role in the explanation of *A*'s preferences in *C*.

---

[7] Here and in what follows, I extend an approach I began in Kitcher 1993. There is kinship between the account I give and that offered by Elliott Sober, first in several articles, and eventually in Sober and Wilson 1998. Sober's version, which focuses on the ordering of preferences, seems to me less general than mine (which considers the intensities as well), but I may be wrong about this.

How this account assembles the elements from our paradigm is clear, I hope. Comments and qualifications are, however, in order. First, this is only a weak form of altruism, because it may consist in a reordering of options that are ranked very low in both preference structures, or even in the adjustment of numerical values assigned to outcomes in a way that leaves the original ordering intact. (Suppose there are two outcomes, ranked by *A* with the values 100 and 0 when there's no effect on *B*; when there is potential for impact on *B*, *A* raises the value of the second to 1; this only makes a difference for ordinal rankings with respect to lotteries that may never arise.) Second, the proposal takes no account of an important kind of altruism where the target of the altruist's reflections is not *B*'s preferences but his *interests*. Third, it does not acknowledge possibilities of mutual adjustment; there is a sense in which one can be incompletely altruistic by failing to take into account the reciprocal altruistic tendencies of another person; further, altruistic desire can often embody a wish that the outcome be achieved in a way that reflects the attempts of the parties to anticipate and to respond to one another's unmodified desires. In what follows, I ignore all these complications.

In fact, I simplify even further. As a normal form for the representation of altruistic desire, we can suppose that, in the solitary situation (C^*), *A* assigns a value v_i to outcome O_i and that, in the case where *B* is present (C), *A* perceives *B* as assigning u_i to O_i. Then the value actually assigned by *A* in *C* can be seen as a weighted average

$$\theta v_i + (1 - \theta)u_i.$$

The value of θ shows the intensity of *A*'s altruism; if θ is 1, then *A* is completely selfish; as θ diminishes, the intensity increases, until, when it takes on the value 0, *A* is so altruistic that she adopts what she takes to be *B*'s preferences as her own.

Altruism is a notion with many dimensions. I distinguish four here. The first is the intensity of altruism just discussed; the second is the *prevalence* of altruism, measured by the range of contexts *C* with respect to which, for a fixed *B*, *A* responds in a weakly altruistic way to *B*; the third is the *extent* of altruism, measured by the range of individuals *B* with respect to which there's some context in which *A* responds to *B* in the weakly altruistic way; finally, *A*'s *empathetic ability* consists in the match between her perception of *B*'s preferences and the actual preferences *B* has. We can easily envisage a four-dimensional space in which people are located by their altruism profiles. Moral philosophers frequently write as if there were a single notion of altruism – the counterpart of egoism – that is morally desirable, but the framework I have offered, simplified though it is, suggests that there may be no single ideal. Indeed, it is not even clear that an ideal moral world would be one in which every person had exactly the same altruism profile.[8]

Much more could be said about this conception of altruism. My announced aim, however, was to outline a perspective on the relation between biology and morality. At this point, that project can begin.

THE RELATION BETWEEN BIOLOGY AND MORALITY

In recent years, some primatologists have made much of the "Machiavellian intelligence" of our evolutionary relatives; in other writings, including the most recent works of Frans De Waal, one finds an emphasis on other-directed behavior.[9] Both sides are right. Chimpanzees, like hominids early and late, are clever enough to play at politics. They also have altruistic dispositions, tendencies that enable them to live together in the first place.

Here are two apparent examples of psychological altruism. The first concerns a young male chimp, Jomeo, who observed an older (and developmentally retarded) female struggling to remove from the climbing frame a tire filled with water. After she gave up, Jomeo went to the frame, solved the problem, and carefully carried the tire to her. In this instance, it is not hard to show that there are no expected benefits for the action – the elderly female, Krom, is not able to do very much for others, she is a low-status member of the troop, the action was unobserved by any others, and so forth. The second example involves a mature female chimpanzee, Little Bee, who spent a period of several months with her partially paralyzed mother, Madam Bee. Throughout the day, Little Bee often lagged far behind other members of the troop as she adjusted her gait to Madam Bee's pace; she would forage for fruit and carefully present half her harvest to her mother. Here again, there

---

[8] One source of complications arises from situations in which *A* can respond to the perceived preferences (needs, interests) of one of two others but not to those of both. Would we prefer that *A* have a consistent tendency to favor one, or that *A* randomize? I am not sure that there's a determinate answer. Or consider the contrast between a low-intensity altruism of broad extent, and a form of altruism focused on a small number of beneficiaries, and prevalent across a wide range of contexts. This is not to deny that there are not some dimensions along which more is better. It is likely that greater empathetic ability is always a good thing.

[9] Ironically, De Waal's first major book was entitled *Chimpanzee Politics*; the view was softened in *Peacemaking among Primates*, and altruism is emphasized in *Good Natured*. I hope that the present essay explains how all three books can cohere, and how the rich variety of De Waal's observations can support a single view of chimpanzee (and early hominid) sociality.

is little possibility of future benefit: Madam Bee was too incapacitated to do anything further for her daughter, and the only primate to observe Little Bee's actions was Jane Goodall.[10]

Are these genuine examples of psychological altruism in the sense I have specified? It seems to me hard to quarrel with the suggestion that the preferences of the apparent altruists are adjusted in light of their attributions of preferences to others: Jomeo sees that Krom wants the tire, Little Bee sees that her mother wants to go more slowly, and, in both instances, their behavior indicates that they have come to want things they otherwise would not have wanted. If altruism is to be dismissed as mere appearance, then the basis, I think, must be the claim that the examples do not conform to the third clause. The suggestion, then, is that the animals modify their preferences as the result of calculation that they will receive certain (selfish) benefits.

What might those benefits be? One thought is that Jomeo and Little Bee expect future aid from the animals they now help; another is that they view their actions as increasing their social standing with third parties in ways that will incline others to help them in the future. On either of these interpretations, Jomeo and Little Bee turn out to be *poor* calculators. To treat Krom and Madam Bee as potential sources of aid is sadly to misjudge the abilities of other organisms; to fail to recognize that there is no audience around to impress is to have a singularly bad appreciation of one's surroundings. I suggest that animals prone to make errors of this kind will find themselves in difficulties in other areas of their lives. To put the point starkly, the hypothetical calculation of future benefits depends on a type of cognitive error that is not only unobserved in the other actions of these animals but is also just the sort of mistake one might expect to be the target of selection.[11]

A different way to criticize the ascription of altruism, one that has worried the most sophisticated defenders of psychological altruism,[12] supposes that the selfish benefit is to be equated with some psychological state, perhaps a warm feeling obtained by giving aid (call it *the glow*) or possibly a condition in which there's no feeling of regret (avoidance of *the pang*). Although our knowledge of chimpanzee psychology is surely inadequate to allow us to rule out the bare possibility that Jomeo and Little Bee are seeking the glow (or absence of the pang), I find this line of criticism relatively untroubling. In the first instance, if the apparent altruists are driven to help others by their

---

[10] The latter example comes from Goodall 1986; the former is from De Waal 1996.

[11] The argument of this paragraph has been improved by discussions with Alex Rosenberg and Roger Sansom.

[12] Sober and Wilson 1998.

quest for the glow (avoiding the pang), then we have to ask why they come to have the pertinent feelings in the context of giving aid. One possibility is that the feeling is founded in anticipation of future benefits or social rewards; but then we return to the more straightforward objections already considered, and the animals are engaged in poor calculation. A different possibility is that the feeling is intricately bound up with helping others, because they respond to the needs of those around them; but that, in effect, is to reintroduce a form of altruism (although the basic response might be an altruistic *emotion* rather than an altruistic preference). Second, if we focus on our own species, we can ask if there are situations in which apparent altruists are able to obtain the emotional reward without actually bringing aid to the other. If there are not, then it reinforces the point just made – the tie between the feeling and the actual relief of others is so tight that it is a mistake to think of the actors giving aid as a *means* to the pertinent feeling; means and ends cannot be separated here. If, on the other hand, there are such situations, then the critic is supposing that cases of apparent altruism are always such that the agents would be indifferent between an outcome in which the glow was received (or the pang avoided) and aid given and an outcome in which the glow was received (or the pang avoided) without giving aid. Not only is there no evidence for that thesis, but it strikes me as highly implausible.

Although more could be said, I think there is a basis for concluding that chimpanzees and hominids share dispositions to form *some* altruistic preferences. How did such dispositions evolve? Perhaps the most primitive such dispositions, shared with relatively asocial animals, were propensities to respond to the needs of the young, and were formed through kin selection.[13] Propensities to form altruistic preferences toward nonrelatives, or more distant kin, were shaped in a different way. Such propensities arose, I suggest, in response to the demands of the coalition game. Weak organisms, struggling for scarce resources, desperately need allies. If they share a tendency to respond to the preferences of others like themselves, then they may be able to participate in stable coalitions and thus improve their prospects. One way to play the coalition game successfully is to have a blind tendency to respond to the preferences of another animal with whom you might engage in cooperative activity.

It is useful to recall here that the coalition game does not succumb to calculation. An animal that makes a friend blindly is likely to do as well as one who tries to figure out what alliances would be most profitable. Maybe better. For, besides the time and energy wasted on fruitless calculations, the

---

[13] Sober and Wilson (1998) discuss this; I offer a scenario in Kitcher 1993.

agent who tries to work out the future benefits may not act as decisively or as constantly as one driven by commitment. Delay and hesitation may undercut the success of the joint venture – and also prove signs that can be used in *reliable* calculations by others, indicators that the indecisive animal is not a valuable member of a coalition.

These speculations are vulnerable to construction of detailed models that specify the pertinent strategies exactly and consider the payoffs in the coalition game. Prima facie, however, whatever the merits of my conjecture that calculation can hurt, it is hard to see how calculation can *help*. If calculation in the context of the coalition game brings no advantage, then we have an answer to the basic question. Propensities to psychological altruism in my sense can evolve because such propensities provide a mechanism for playing the coalition game that is no worse than the political strategies – so-called Machiavellian intelligence – we might otherwise ascribe to our ancestors.

This yields a reconciliation of Darwin and Hume. Psychological altruism, as I understand it, refines the Humean notion of fellow feeling. I do not pretend that the case is conclusive, but there are, I believe, grounds for supposing that dispositions to psychological altruism, toward kin and non-kin, are part of our evolutionary heritage.

THE FRAGILITY OF PSYCHOLOGICAL ALTRUISM
AND PEACEMAKING STRATEGIES

Those dispositions are fragile and partial. Recall my multidimensional conception of altruism. In the social world of chimpanzees and hominids, we might expect the nested social structure – coalitions and subcoalitions – to reflect differences in the intensity and prevalence of dispositions to altruism. Perhaps any pair of organisms in a band will share some contexts in which each responds to the perceived preferences of the other, but there will be considerable differences in intensity and prevalence if dyads are chosen at random from the group. From the close friends always observed together to the animals who unite only in the face of an outside threat, there is a complex spectrum of cases.[14] Yet even where a pair has formed an apparently stable coalition, the prospect of especially large rewards for defection can strain the relationship. Altruism seems incompletely prevalent.

[14] Goodall discusses several instances of chimpanzees who are inseparable; both she and De Waal provide examples of animals with more distant relationships.

De Waal's observations of the Anthem chimpanzee colony provide a striking case. Over a period of years, an older male, Yeroen, had been dominant in the group, using alliances with females to defeat challenges to his authority. Those challenges came from two younger males, Luit and Nikkie, who acted as a team. As their powers increased and Yeroen's declined, the balance tilted in favor of the challengers. After their victory, however, there were several important shifts. First, Luit, the stronger of the two rebels, cultivated alliances with the females, and deprived Nikkie of opportunities for mating. This initiated a time of great instability in the group, marked by successive alliances between Nikkie and Yeroen, Luit and Yeroen, and individual males with groups of females, before, one night, Yeroen and Nikkie ganged up on Luit, biting him so badly that he died on the operating table.[15]

Here my main concern is not with the sad ending but with the transition that caught De Waal's attention, even before he knew how things would turn out. As he makes plain, De Waal found Luit's defection from his coalition with Nikkie profoundly surprising, and he interpreted it in terms of chimpanzee politics, the Machiavellian calculation of advantage. That interpretation strikes me as correct, but it should not lead us to suppose that Luit had been calculating all along. From the perspective I offer, there is no difficulty in supposing that an incompletely pervasive disposition to psychological altruism can be exposed when a context arises in which an animal has an opportunity to secure a very large benefit. That, I suggest, was Luit's situation once Yeroen had been dethroned. The prize of sole dominance was simply too large, and the disposition to respond altruistically to Nikkie was no match for it.

A second aspect of De Waal's study deserves note. During the time of instability, when the usual coalitional bonds were broken on a daily basis, the animals spent more and more time in intimate contact with one another. After engaging in agonistic interactions, Luit and Nikkie would nervously approach one another and settle in for an unusually extended period of grooming. I think De Waal is correct in interpreting this as a peacemaking strategy that held together a community in considerable danger of falling apart.[16]

This example highlights features found in every chimpanzee (or bonobo) group that has been studied. Coalitions sometimes endure over long periods of time, but, in almost all cases, their members sometimes fail to do what their allies expect of them. A constant feature of the maintenance of the social structure, therefore, is the reassurance provided by intimate interactions. As I

[15] De Waal narrates this story brilliantly, giving the first part in De Waal 1982, the second in De Waal 1989.
[16] See De Waal 1989.

suggested earlier, the amount of time spent in grooming is incomprehensible if one thinks in terms of the hygienic benefits; simply far too much time and energy are spent. By contrast, once one realizes the importance of holding together a complex of coalitions and subcoalitions, one can see the work that is being done.

From a psychological point of view, the dispositions to altruism are insufficiently pervasive. They allow for forms of behavior that threaten the stability of the social order. Chimpanzees thus live in fragile societies, held together in part by the investment of time in peacemaking, most notably in social grooming. Their bands cannot be bigger than they are, because the combination of altruistic propensities and peacekeeping support would be inadequate to maintain a larger society. Their opportunities for developing a broader range of cooperative ventures are similarly reduced. They live, in short, in a state of nature.

Hobbes, then, was partly right. In declaring the condition of war to be one in which the threat of violence is always present – by analogy with the English weather that always portends rain – he painted a picture that fits chimpanzee, and probably hominid, society very well.[17] The social fabric is in constant danger of being torn, and it requires considerable effort to patch it together. But Hobbes was only partly right. For he supposed that the state of nature was an expression of a strict tendency to psychological egoism. On the contrary, I suggest, the Hobbesian threat arises only because we have enough of the Humean propensity for altruism to form a minimal society in the first place. Chimpanzees are limited Humean altruists who face Hobbes's problem.

Hominids somehow learned to do better. Their – our – progress is indicated in the much larger groups in which we can live and in the variety of cooperative ventures we can undertake. How was that possible? I conclude by offering an answer, backed by all-too-little evidence – *through the evolution of a capacity for normative guidance*.

THE EVOLUTION OF A CAPACITY FOR NORMATIVE GUIDANCE

Chimpanzees and bonobos experience internal struggles as their tendencies to psychological altruism conflict with their other preferences. Those struggles become visible sometimes when one animal has obtained a valued resource and another comes a-begging. The fortunate owner will extend the branch (or whatever) toward the supplicant, at the same time averting his face. The

17 Here I'm indebted to Michael Williams.

strain of the pose reveals the warring desires, just as the dieter's salivation as he firmly marches past the wonderful smells from the restaurant tell outsiders what is going on. Chimpanzees, and our hominid ancestors, are/were vulnerable to the internal melee of competing suggestions, and the voices that shouted loudest are/were sometimes socially destabilizing. Their psychological anarchy produced social disorder and involved them in elaborate and time-consuming bouts of peacemaking.

A decisive transition in human prehistory, possibly one that accompanied and helped shape the emergence of language, was the acquisition of a capacity for reinforcing the altruistic tendencies, a capacity that was able sometimes to forestall the direction of behavior by preferences for selfish benefits. With respect to individuals with whom a hominid interacted on a daily basis, that capacity decreased the frequency of social rupture, so that the wasteful business of repairing broken social bonds was required less frequently. Liberated from constant peacemaking, hominids could explore a wider range of cooperative ventures – here we return, at last, to the benefits discerned in the original perspective of optional games – and they could even enter into projects with others who were only encountered irregularly.[18]

The crucial change is the ability for self-governance according to a system of rules. Let me state, baldly, some of the steps that may have been important in the emergence of that ability. I begin with the development of systems of punishment.

Plainly, even in groups in which there is no genuine punishment, animals engage in agonistic encounters. Let us ask, then, what conditions are required to transform the agonistic interactions in such groups into real punishment. In the initial state, I suppose, the adverse reaction of one animal to the behavior of another is quite uncoordinated with the behavior of others; sometimes others may intervene to help the "victim," on other occasions not. A first step in the direction of punishment seems to be that other members of the group, even those who may be allies of the threatened animal, should not intervene. Thus we can envisage populations in which there is a regular pattern; with few exceptions, aggression in contexts of particular types does not cause the allies of the aggressor's target to rally round – the allies "let" the aggression go forward. Next we can imagine that the mere regularity is coupled with an expectation, shared by the organisms in the population, that others will not interfere in such contexts. Further, this expectation may lead to no resistance

---

[18] Even in the Upper Paleolithic, hominids seem to have fashioned tools from materials only available at distant sites; by the early Neolithic, there is evidence for long-distance trade. See Postgate 1992, chap. 11.

on the part of the target; the animal picked out merely suffers what happens. Yet another refinement would be the existence of a regularity concerning the animals who carry out the aggression: perhaps they are animals who bear a particular relation to the context, perhaps they play a particular social role. Finally, we can suppose that there is an expectation about the identities of the animals who initiate aggression. At this last stage, we seem to have reached the systems of punishment found in contemporary human societies (and in societies for which we have historical records).

I do not want to claim that the evolution of punishment necessarily followed the steps just envisaged; nor do I want to specify a point at which "real" punishment is present; nor shall I offer any detailed account of the reasons why any hominid lineage might have undergone this sequence of steps. Firm views on the last issue ought to be grounded in precise models of the advantages of moving from one stage to the next, and while I have an outline of how such models might be developed – roughly in terms of the advantages in opportunities for cooperative activity without costly signaling among organisms that have moved from one stage to its successor – showing how that intuitive idea can be elaborated within the joint theory of biological and cultural evolution that I favor would require substantial work. Finally, it should be evident that the early stages of the envisaged sequence can be managed without language and that later steps would, at the very least, be facilitated by the prior acquisition of linguistic skills, but I am not going to link this sequence in any definite way to the evolution of language. For present purposes, ideas about systems of punishment are relevant insofar as they illuminate questions about self-governance. So, setting the important issues I have noted on one side, let me imagine that our hominids have acquired a full-fledged system of punishment, corresponding to the last stage I delineated, and that they are able to formulate their expectations in language.

Our evolving hominids thus can entertain and believe propositions of the following forms:

> If, in C a P does W, then a J will typically respond by doing S to that P. [Here the letters stand for types of organisms, acts, and situations: think of C as standing for the context, P for the perpetrator, W the wrongdoing, J the judge, and S as the sentence.]
> On such occasions, other members of the group typically won't interfere with J's doing S.
> On such occasions, the P typically won't resist the J's doing S.

Now we imagine that hominids with self-governance and those without it differ in that the former, but not the latter, have a mechanism that tends to give rise to a reactive emotion (an unpleasant emotion) first on occasions on which they have performed the wrong in the context (done *W* in *C*), and subsequently on occasions on which they are in the context and have formed a disposition to prefer doing the action marked out for sanction (when they come to prefer doing *W* in *C*). I suppose further that the consequence of feeling the reactive emotion *after* doing the action consists in an enhanced disposition to submit to the sentence, and that the consequence of feeling the reactive emotion *before* carrying out the action (when the hominid comes to form the preference in prospect) is to diminish the strength of the preference for the action.[19]

There are two kinds of cases we need to consider. In one, the agent is genuinely torn; there are conflicting dispositions, one ranking one option as preferable, the other reversing that ranking, as I have suggested for the chimpanzees torn between altruistic responses and the enticement of large selfish rewards (recall Luit abandoning his alliance with Nikkie). The new mechanism serves in such instances as an instrument for sorting out the internal melee, although its performance need by no means be perfect. In the alternative scenario, there is no conflict of dispositions, and the role of the unpleasant emotion is to weaken or reverse the preference for doing *W*. In both instances, the net effect of the mechanism is a tendency to avoid the actions that lead to trouble. To the extent that the desires are inhibited, our hominids do not incur punishments they would otherwise likely have received. (Of course, the account I have offered must recognize that punishment is not the inevitable consequence of action, for the action might go undetected; there are interesting questions about whether the hominids could evolve mechanisms for "overrepresenting" the chances that they will be caught, thus leading to the occurrence of the unpleasant emotion in circumstances where the objective likelihood of detection is low.)

The story so far has envisaged a transition from hominids who sometimes transgress the punishment regularities of their groups and are punished for doing so to hominids whose psychology contains a mechanism that operates prospectively to decrease the probability of transgression. It seems likely that an evolutionary understanding of that transition might be gained along the general lines I indicated for thinking about the emergence of systems of punishment – but, as always, detailed models are needed and I am not going to provide them here. What concerns me is the character of the final state. It is tempting to think of this as consisting of a rather abstract inhibition

---

[19] My proposal here is akin to that offered by Allan Gibbard (1990), although, unlike him, I am not concerned to distinguish *being guided by a norm* from *being in the grip of a norm*.

device: the aversive emotion will be generated in certain contexts with respect to certain prospective actions, and that emotion will weaken the preference for those actions; but what these actions and contexts are is a matter for different societies to fill in (in much the way, perhaps, that exposure to a particular language fills in a child's grasp of universal grammar). So we might conceive of the mechanism as completely open to whatever content the hominid society supplies, as if *any* social rule could take effect with equal ease. A different thought is that the hominid system of normative governance is biased toward certain types of rules; in the simplest (maybe not the most plausible) version, one might even suppose that there are built-in reactions to some actions whether or not they are explicitly forbidden by society (this is a line that could elaborate the sociobiological insistence on a biological basis for incest avoidance). If the governance mechanism is conceived as relatively plastic, then, to the extent that common reactions are found in all human societies, those will be explained by supposing that societies whose rules failed to set up those reactions had a tendency (possibly explicable, possibly accidental) to die out. On the sociobiological alternative, the reactions are universal because of features of human nature, whatever the expressed rules of various societies may be. In my judgment, we do not really know how to resolve the issues here, but I think there are enough examples of cross-cultural variation in reactions to demonstrate that large parts of the normative governance system are "filled in" by the ambient culture.

If, as I have suggested, the system of normative guidance substitutes for inefficient and time-consuming strategies of peacemaking, then it is easy to make educated guesses about the content of the rules that societies would attempt to inculcate. Recall that the social problems to which peacemaking is addressed are lapses from cooperative behavior, where the terms of cooperation are set by the coalitional and subcoalitional structures present in the group. Thus we should expect rules that make the forms of coalitions and subcoalitions visible, and that enjoin loyalty. Further, because the occasions on which social tension is most threatening involve intragroup violence and the opportunities for mating, we might anticipate that the rules should specify when violence is to be prohibited and which pairs of hominids may engage in sexual relations. We might conclude, in short, that the social rules should embody the "elementary structures of kinship," that they should pronounce on acts of violence, and that they should include marriage rules. It is no accident, I think, that such rules constitute the core of the normative systems of those groups that live in ways closest to those experienced by our hominid ancestors.

I can now make my hypothesis more concrete. A decisive step in hominid evolution consisted in the acquisition of a capacity for normative guidance, and the "filling in" of that capacity with rules of group loyalty, including explicit proscription of violence across a range of contexts and explicit rules about marriage and mating. Hominid groups that were able to achieve this system – call it "protomorality" – were able to engage in the older repertoire of cooperative ventures with greater efficiency, and were also able to undertake new cooperative projects. Their surviving descendants are linked to them through a sprawling genealogical tree, along whose branches different systems of socialization have proliferated, introducing psychological differences among (and within) various cultural groups, as well as different adumbrations and revisions of the initial set of prescriptions and prohibitions. The primary force in the dynamic of moral change has been the differential ability of groups with different moral codes and systems of moral training to survive, to spread their views to others, and to found new groups in which their ideas would be accepted. Out of the process have emerged the moral systems of contemporary societies, including those of the affluent world.

All this is conjecture. How could it be supported? Only, I think, through a massive compilation of psychological, sociological, anthropological, and historical facts that are brought together and illuminated from the perspective I have offered – by carrying out, in short, the strategy Darwin used so successfully in the *Origin*. Even if I knew how to do it (which I do not), that would be impossible here. I can only offer a brief indication of one source of evidence.

What light does my story shed on the earliest moral systems in the historical record? The most famous system of moral rules in the Western tradition derives from writings that recapitulate much older laws and precepts. From the beginning of the second millennium, clay tablets record fragments of the laws that governed Mesopotamian societies. The preambles to these "codes" constantly emphasize the idea that the lawgiver brings peace and resolution of conflicts; the law is seen as a method of transcending a social life in which brute force prevails and the strong oppress the weak. Further, it is evident that the tablets and stelae that have come down to us do not offer any complete account of the laws in force. They are sets of amendments to a body of existing law, revisions and extensions that address problems that seem to have arisen in the creation of social order. I do not think it is fanciful to take these "codes" as representing a multistage process of development of the social rules that extends back to the dawn of writing and beyond. The fragmentary character of the codes is immediately obvious. Provisions are made for very particular types of occurrence – whether a "senior" strikes the daughter of another "senior" and causes a miscarriage, whether an ox gores a passerby, whether a woman crushes the testicle(s) of a man who is fighting her husband.[20] I interpret this particularity as pointing to a practice of responding to the new kinds of troubles that emerged in a newly complex society.

By the time of Hammurabi, people had been domesticating animals and engaging in agriculture in Mesopotamia for at least five thousand years. There had been settlements of significant size in neighboring regions (at Çatal Hüyük, and at Jericho), although nothing on the scale of Uruk or Babylon. The Neolithic pastoralists and farmers of the region had worked out rules for restraining violence, protecting the fruits of their labors, and organizing sexual relations. But as they were integrated into larger units in a world dependent on social coordination to supply adequate irrigation, new issues arose – how are measures to be standardized, how does one ensure that land is properly used, how are the public canals and dikes to be maintained? The codes we have lavish great detail on these questions, as well as addressing the various kinds of violence and sexual relations that emerged from the social friction of large numbers of people occupying a relatively small space. They occur against the background of a general understanding of the ways in which violence is to be contained, sexual relations regulated, and property protected.

The first two are the original contexts in which normative guidance served to transcend the peacemaking activities of early hominids, the contexts in which previous societies threatened to break down. The last becomes important, I believe, when the amplification of cooperative ventures leads to division of labor and the possibility of surplus. Hammurabi, let alone the authors of the Pentateuch or Socrates, comes very late in the history of human morality, refining precepts and moral ideas that had been worked out as prehistorical members of our species used their capacity for normative guidance to achieve more stable social structures (and more stable psychological lives).

I began by promising an alternative to two polar views, one that would give a role to our altruistic tendencies (and to biological evolution) and also a role to the evolution of a culture that shapes our minds and behavior. I hope it is apparent how the story I have told fulfills that promise – but I fear that my advertisement of even a *partial* defense may seem an unwarranted overstatement.

---

[20] Both the latter cases appear in the Hebrew Bible, plainly recapitulating the Mesopotamian traditions.

REFERENCES

Axelrod, R. 1984. *The Evolution of Cooperation.* New York: Basic Books.

Batali, J., and Kitcher, P. 1995. Evolution of Altruism in Optional and Compulsory Games. *Journal of Theoretical Biology* 175: 161–171.

De Waal, F. 1982. *Chimpanzee Politics: Power and Sex amongst the Apes.* New York: Harper and Row.

De Waal, F. 1989. *Peacemaking among Primates.* Cambridge, Mass.: Harvard University Press.

De Waal, F. 1996. *Good Natured: The Origins of Right and Wrong in Humans and Other Animals.* Cambridge, Mass.: Harvard University Press.

Gibbard, A. 1990. *Wise Choices, Apt Feelings.* Cambridge, Mass.: Harvard University Press.

Goodall, J. 1986. *The Chimpanzees of Gombe: Patterns of Behavior.* Cambridge, Mass.: Belknap Press of the Harvard University Press.

Hamilton, W. D. 1971. The Genetical Evolution of Social Behavior. In *Group Selection,* ed. G. C. Williams. Chicago: Aldine.

Kitcher, P. 1993. The Evolution of Human Altruism. *Journal of Philosophy* 10: 497–516.

Maryanski, A., and Turner, J. H. 1992. *The Social Cage: Human Nature and the Evolution of Society.* Stanford: Stanford University Press.

Maynard-Smith, J. 1982. *Evolution and the Theory of Games.* Cambridge: Cambridge University Press.

Postgate, J. N. 1992. *Early Mesopotamia: Society and Economy at the Dawn of History.* London: Routledge.

Sober, E., and Wilson, D. S. 1998. *Unto Others: The Evolution of Altruism.* Cambridge, Mass.: Harvard University Press.

Smuts, B. B., Cheney, D. L., Seyfarth, R. M., Wrangham, R. W., and Struhsaker, T. T. (eds.). *Primate Societies.* Chicago: University of Chicago Press.

Trivers, R. 1971. The Evolution of Reciprocal Altruism. *Q. Rev. Biol.* 46: 35–57.

Williams, G. C. (ed.). 1971. *Group Selection.* Chicago: Aldine.

Wrangham, R. 1987. The Evolution of Social Structure. In *Primate Societies,* ed. B. B. Smuts et al., 282–296. Chicago: University of Chicago Press.

10

Will Genomics Do More for Metaphysics Than Locke?

ALEX ROSENBERG

> Origin of man now solved. He who understands baboon would do more for metaphysics than Locke.
>
> Darwin, *Notebooks*

THE EVOLUTION OF HUMAN BEHAVIOR AND "JUST-SO STORIES"

Darwin's claim is probably guilty of pardonable exaggeration. After all he did not prove the origin of man, and Locke's greatest contributions were to political philosophy, not metaphysics. But it may turn out that Darwin's twentieth-century grandchild, genomics, vindicates this claim with respect to both metaphysics and political philosophy. Here I focus on the latter claim alone, however.

From the year that William Hamilton first introduced the concept of inclusive fitness and the mechanism of kin selection, biologists, psychologists, game theorists, philosophers, and others have been adding details to answer the question of how altruism is possible as a biological disposition. We now have a fairly well-articulated story of how we *could have* gotten from there, nature red in tooth and claw, to here, an almost universal commitment to morality. That is, there is now a scenario showing how a lineage of organisms selected for maximizing genetic representation in subsequent generations could come eventually to be composed of cooperating creatures. Establishing this bare possibility was an important turning point for biological anthropology, for human sociobiology, and for evolutionary psychology. Prior to Hamilton's breakthrough it was intellectually permissible to write off Darwinism as irrelevant to distinctively human behavior and human institutions. The unchecked contempt with which defenders of the autonomy of the social from the biological operated in their attacks on naturalistic approaches to social processes was both breathtaking and without effective rejoinder (cf. Sahlins 1974). The problem of how even to reconcile the theory of natural selection with the possibility of cooperative institutions was so grave that E. O. Wilson insisted that Camus was wrong: it was not suicide that is the only philosophical question, but rather altruism (Wilson 1975, p. 3).

The major components of the research program, the models and simulations, the comparative ethology, are well known. Once Hamilton showed that inclusive fitness maximization favors the emergence of altruism toward offspring, a virtual riot of ethological activity began to identify previously known cases of offspring care as kin-selected and to uncover new examples of it. Once Hamilton was joined by Axelrod in identifying circumstances under which reciprocal altruism between genetically unrelated beings would be selected for, the community of game theorists began to make common cause with evolutionary biologists in the discovery of games in which the cooperative solution is a Nash equilibrium. This led in turn to the development of models of evolutionary dynamics for iterated games like cut-the-cake, ultimatum, and hawk versus dove that show how a disposition toward equal shares, private property, and other norms among genetically unrelated beings may be selected for. An independent line of inquiry at the intersection of psychology and game theory developed an account of emotions suggesting that they too may have been selected for in order to solve problems of credible commitment and threat in the natural selection of optimal strategies in single games.

But in a sense all this beautiful research remains what Gould and Lewontin (1979) once characterized as a "just-so story." Can we convert the "how possible" explanation of human sociality into a testable and tested chronology of the actual evolutionary origins of cooperation as an adaptation? Well, what are our resources? They seem slim. After all, the relevant phenotypes, if any, are not among the hard parts preserved in the fossil record. It is not just that no missing links or transitional forms have turned up; there seem not to be any links. There is, of course, comparative ethology, neurophysiology, and neuroanatomy. But, at most, these provide the data from which we can reverse-engineer our way into – well, into just-so stories, hypotheses among which we cannot choose on the basis of independent evidence.

Only one evidential source stands a chance of doing any better: genomics. In this chapter I want to explore what genomics can reveal about the actual evolution of human cooperation, when combined with phylogeny, comparative ethology, neurophysiology and neuroanatomy, and paleontology. To see its potential however, first, let's consider what it can show us about recent human prehistory. This can give us an idea of how genomics can turn questions hitherto supposed to be purely speculative into matters open to testable answers.

GENOMICS AND THE EXPLANATION OF THE EVOLUTION
OF HUMAN BEHAVIOR

By genomics I mean the comparative and often computational study of the nucleotide sequences and the functional organization of the human genome and the genomes of many other species of animals, plants, and fungi. The Human Genome Project has already given us a first draft of the 3 billion base pair DNA sequence of the human genome. It has so far given us a little more information about the human genome. For instance, it appears that even more of it is "junk" DNA than molecular biologists have thought; "junk" DNA has no role in development or normal human function and is just along for the ride, so to speak. And it now appears that there are only about 30,000 to 60,000 genes in our genome, which makes it little more than twice or four times the size of the fruit fly's genome. But at an accelerating rate genomics – the comparative study of the human and the DNA sequences of other organisms – will begin to give us the sort of detailed information about our genomes we never dreamed of, and will give it to us as the result of methods we can automate and turn over to computers. Learning about our genomes and their protein products will cease to require genius, and at most demand ingenuity. Learning exactly which DNA sequences among the 3 billion nucleotide bases express genes and which genes they express is a matter of "annotation" of the DNA sequence the Human Genome Project has provided. Even before the whole sequence came into our hands, comparative genomics was illuminating large tracts of history about which only informed speculation had hitherto been possible.

We are inclined to think of history as having begun when written records did, about 3,000 years ago in the Near East, and 1,000 years later in Mesoamerica. But DNA sequence data already in hand extend our knowledge of the general lines of human history so far back as to turn the Inca empire, the fall of Rome, the building of the Great Wall of China, or the building of Sumerian Ur into matters of recent history. DNA sequence data can answer detailed perennial questions about human origins and prehistory that have hitherto been the domain of pure speculation. Like the bar code on a can of beans on a supermarket shelf, our DNA sequences are labels from which we can read off date and place of manufacture not just in geological time, but over the past 200,000 years with resolving power that already approaches only a few thousand years, just beyond the reach of carbon 14 dating. Seeing how fine-grained is the resolving power of the genetic bar code in these cases should give us some confidence it can answer countless other questions hitherto beyond the reach of nonspeculative answers. But to see this requires a little of the science of DNA sequences.

Individuals inherit their cells' mitochondria and the genes they contain only from their mothers, because the mitochondrion genes are not in the nucleus of any cell – somatic or germ line – and so do not make it into the sperm, which contains only DNA from the nucleus. But because mitochondria are in every cell, they are in every ovum, and so in every ovum fertilized by a sperm. By contrast, all males and everyone else who has a Y chromosome inherits it from his (or her) father (the parenthesized here accommodates the rare XXY females). Mitochondrial genes' DNA (mtDNA) and Y-chromosome DNA can be sequenced. Because individuals differ from one another in gene sequence, it is easy to order a sample of individuals for greater and lesser similarity in DNA sequences – whether in the nucleus or the mitochondrion. The more similar the sequences, the more closely related two people are. Given an ordering of similarity in mtDNA and Y-chromosome DNA among people living today, and comparing it to some mtDNA and Y-chromosome DNA sequence in another species whose age is known, geneticists can work backward to identify an mtDNA or Y-chromosome sequence from which all contemporary sequences must have mutated and descended; in effect they can draw a family tree of all the main lines of descent among mtDNA or Y-chromosome sequences, and they can date the age of various branches in this family-tree. mtDNA sequence data were available much earlier than Y-chromosome data; they led to the conclusion that every human being now living is descended from one particular woman living in eastern Africa – current Kenya and/or Tanzania, approximately 144,000 years ago. She alone, of all women then alive, has had an unbroken line of female descendants from that day to this. Every other woman has had at least one generation of all male descendants, and so her mitochondrial sequences have become extinct.

Moreover, the narrowness in sequence variation among extant people reveals that we are ten times more similar to each other in sequence data than, for example, chimpanzees are similar to one another in sequence data. It can also be established that this woman, called "Eve" by biological anthropologists, lived among a relatively small number of *Homo sapiens*, who must have gone through some sort of evolutionary bottleneck – that is, most of our ancestors were killed off at some point in the recent past. As a result there were only about 2,000 (±1,000) women altogether alive at the time Eve lived. Subsequent sequencing of the Y chromosome has confirmed these conclusions. Indeed, as more and more sequence data come in, about single nucleotide polymorphisms and microsatellite loci, the conclusion has become inescapable (in spite of Chinese reluctance to accept it) that all present *Homo sapiens* are descended from this one African Eve and a relatively small number (about a dozen) of African Adams alive at the same time as Eve.[1] This explains why intraracial gene sequence differences are larger than interracial ones, why polygenetically (many gene) coded traits have not had sufficient time to assort into separate lineages, and thus why race is not a biologically significant explanatory concept. The genetic similarity among humans suggests further that the obvious visible differences among us in skin color, hair color, facial characteristics, and the like are both of relatively recent origin and are most probably the result not of natural but of sexual selection.

The data that suggest to some that humankind went through a narrow bottleneck before expanding rapidly across the entire surface of the earth are also consistent with this particular band figuring as a founder population, which survived when other *Homo sapiens* groups did not because of some intergroup fitness difference, such as bestowed by intragroup cooperation, for example. This of course is a bit of data that tend to support at least slightly a theory like Sober and Wilson's account (1998) of the origins of cooperation.

Besides telling us where and when we started from, following our differences in more and more available DNA sequences, geneticists have traced the details of early human migration out of East Africa both into western and southern Africa, and northward, dating the arrival of *Homo sapiens* on each of the continents to within a few thousand years, and explaining in some detail the peopling of Micronesia, Melanesia, and Polynesia within the past 6,000 years (see Cann 2001).

And beyond chronology, sequence data provide other startlingly detailed revelations about matters of prehistorical narrative hitherto thought forever beyond answers. For example, consider the question that concerned novelists like Auel and Goulding, and many others: what happened to the Neanderthals? Well, Neanderthal DNA is available in bones from the Neander valley in Germany. By comparing mtDNA and the ALU gene sequence – a bit of junk DNA sequence that repeats a distinctive number of times in chimp, *Homo*

---

[1] For an introduction to the African "Eve" hypothesis and supporting data, see Boyd and Silk 2000 (pp. 477–483), Hedges 2000, and articles there cited, especially Stoneking and Soodyall 1996. For Y-chromosome sequence confirmation and amplification, see Renfrew, Foster, and Hurles 2000 (and papers there cited), and Stumpf and Goldstein 2001.

sapiens, and Neanderthal DNA – it can be shown that these three lines of descent don't share these genes at all, as they would have to if there were any interbreeding among them. This is not surprising in the case of chimpanzees and *Homo sapiens,* of course. But that there was no interbreeding between our species and Neanderthal at all is very significant. That means that either *Homo sapiens* killed off the Neanderthal or gave them all a fatal disease, or otherwise outcompeted them in a common environment. Probably, Cro-Magnon outcompeted them, because there is archaeological evidence that both populations existed side by side in Europe over many thousands of years (Boyd and Silk, 2000, pp. 484–485; *Science,* 11 July 1997; Gibbons 2001a). Similarly, the absence of any non-African Y-chromosome sequences among 12,000 Asian males from 163 different populations shows that the migrants out of Africa replaced any earlier Asian populations and did not interbreed with them either (Ke et al. 2001).

Further research will employ DNA sequence data to uncover the detailed narrative of events we never dreamed of reconstructing and of other events our nongenetic records have misrepresented to us. For example, consider the origin of agriculture in Europe about 10,000 years ago. How did it happen? There is some archaeological evidence to show that farming spread from the Near East northward and westward in Europe. But how? By cultural evolution one might presume: farming must have spread as people in one European valley noticed the success of those farming in the next valley to the southeast and copied their discovery. Others have held that the farmers came out of the Near East, and like the Cro-Magnon's outcompeting or extirpating Neanderthal, displaced – pushed out or decimated – local populations, took over their territory, and thus expanded the farming regions. Which hypothesis is right is not a question we could ever have expected to answer because these events took place before any *recorded* history, indeed before writing!

But recent studies, first of mtDNA, and now of Y-chromosome sequence differences in contemporary Near Eastern and European populations, substantiate the latter scenario, the so-called demic-defusion model, a euphemism for the displacement of one whole population by another. MtDNA and Y-chromosome sequence data show that the earliest migration from the Near East into Europe occurred about 45,000 years ago, and its descendants now account for only about 7 percent of contemporary European mtDNA, but earliest immigrants provide twice that proportion of mtDNA among the isolated Basque, Irish, and Norwegian populations, and only half that frequency in Mediterranean populations. The next wave of migration about 26,000 years ago provided about 25 percent of current mtDNA in Europe, while the third wave 15,000 years ago accounts for about 36 percent of contemporary

European mtDNA. Agriculture arrived with a diffusion from the Middle East about 9,000 years ago, and despite their recent arrival the mtDNA sequences these immigrants brought with them account for 23 percent of the mtDNAs of current European populations, 50 percent when we exclude the extreme Basque, Irish, and Scandinavian populations. And this wave of migration provides mtDNA and Y-chromosome DNA sequences in a "cline" – a gradient of change in proportions – that moves in the direction from southeast to northwest (Richards et al. 2000). What the sequence data tell us is that Near Eastern populations displaced indigenous ones year after year in wider and wider arcs of expansion from the Middle East, either driving them west eventually to the extremities of the European continent or killing them off so that the only survivors of the original population of Europe were those inhabiting agriculturally marginal territories.

The question arises then, why didn't the earlier inhabitants acquire farming either independently or by imitation of their neighbors' practices? Surely there is no gene for farming that they lacked. Did farming and the social organization it produced make the Near Easterners that much more formidable than the hunter-gatherers? If so, why? Further thought about this displacement should at least enable theorists of the evolution of cooperation among hunter-gatherer egalitarians to set some constraints on their models. The payoffs to cooperation cannot be so strong as to prevent defeat by less egalitarian groups with storable commodities.

More recent population events, besides revealing who settled Melanesia, Micronesia, Polynesia, and the Western Hemisphere and when they did so, will tell us who arrived later, what groups went back the other way to settle Madagascar (where mtDNA sequences are quite different from mainland African mtDNA)[2] and why the current residents of the Andaman Islands east of the Indian mainland have mtDNA sequences far closer to those of East Africans than even the inhabitants of their neighboring islands or the Indian subcontinent.

Nonhuman DNA sequence data will be able to tell us more about human prehistory. Sequencing the domesticated plants and animals and their extant undomesticated relatives can tell us where and when hunting and gathering first gave permanent way to farming, and thus to the beginnings of hierarchal social, political, and cultural institutions. And they can date these events well before or with much greater accuracy than does the archaeological evidence now available. In fact, what DNA sequence research thus far has shown is that both wheat and cattle were probably domesticated at least twice independently

[2] Cf. Gibbons 2001b, and papers cited therein.

and at roughly the same time. Among the earliest domesticated cereals is emmer wheat, which however reflects two different sequences that diverged 2 million years ago, one traceable back only to southern and central Europe, including Italy, the Balkans, and Turkey, while the other is ubiquitous to all regions of emmer cultivation. This suggests a double expansion from domestication in the Middle East. There are two distinct types of cattle – the humped breeds of India and the humpless ones of Africa and Europe. They were both domesticated 2,000 years after wheat, but their DNA sequences are sufficiently different to support the hypothesis of separate domestication (T. A. Brown et al. 1998; Turner et al. 1998).

Multiple initial domestication is, of course, suggestive. Recall Dennett's point: "If you wanna live, you gotta eat. In chess, when there is only one way of staving off disaster, it is called a forced move" (Dennett 1995, p. 128). Might domestication have been something approaching a forced move by populations that had depleted so much of the megafauna of their environment, that hunting was no longer a viable option? If domestication were fortuitous serendipity, a "neat trick" in contrast to Dennett's "forced move," one would have expected a single initial experiment with no evidence of distinct DNA sequence variation in the resulting descendants. Two independent bovine domestication events happening in two different and independent places twice in a couple of thousand years may not give us enough grounds to conclude that for human survival it was the only way to go. But multiple independent and roughly simultaneous domestications around the world would tend to suggest it is a forced move. Of course, the distinction between "forced moves" and "neat tricks" is not a neat one. When one of a number of available solutions to a design problem is vastly superior to the others, and is in fact taken up by a lineage in design space so that it eventually becomes fixed in the whole species, is it a forced move or a neat trick? Let's call it, without Dennett's permission, a forced move.

Were we to be able to pin down the emergence of sociality to a number of similarly independent scenarios, we might add further strength to the notion that cooperation is strongly selected for. There are problems here, however, that we need immediately to face. First, if the just-so stories have it right, cooperation, reciprocal altruism, a sense of justice as equal division, and the emotions and the norms that enforce and express these institutions long antedate agriculture. Moreover, if anything, we should expect agriculture to begin to provide an environment in which many of the dispositions we seek to explain may cease to be selected for. Once agriculture kicks in, the inclination to equal effort and equal shares becomes much less adaptive for individual survival. Storable commodities and capital investment emerge, and

185

the payoffs to cooperating, sharing, reciprocation, defecting, hoarding, and free-riding, not to say domination, become quite different, hard to model, and probably produce unstable equilibria. Second and relatedly, the mitochondrial DNA sequences strongly suggest that sometime at or before 144,000 years ago, there was a bottleneck through which *Homo sapiens* came. This was long before the advent of agriculture, and presumably cooperation was already well established at that point. If *Homo sapiens* is the sole species to show the degree of cooperation we seek to explain the emergence of, there is no question of multiple independent simultaneous candidates for the designation of evolutionary "forced move." At most there is the "neat trick," which could have emerged and persisted without specific genetic underpinning. But, of course, if other hominids found themselves forced into a high degree of cooperation in order to survive, and we had evidence of this, the forced-move hypothesis would gather some strength. But even if sociality is written into the gene sequence, a tendentious assumption yet to be discussed, it is obvious from gene sequence data that these other hominids left no representatives for us to sequence and compare. Or did they?

Recall that DNA has been extracted from Neanderthal bones upward of 40,000 years old. This work is part of a new subdivision of biological anthropology that styles itself the study of *ancient DNA*. Quantities of DNA to be found in burial ground bones, around cave and campfire detritus (and coprolites for that matter), or in fossil skulls are minuscule in quantity; proportions of the full sequence are low, and no particular portion – say, functional genes as opposed to junk DNA – is preferentially preserved. Nevertheless, the prospects of worthwhile data are not entirely unfavorable. The optimism here as elsewhere in the genetic revolution is in the power of a molecular process, PCR, or polymerase chain reaction, for the amplification of DNA. This process employs a reagent that can catalyze the amplification (reproduction) of a single nucleotide sequence of any length into a million copies in only thirty rounds of replication. This means if a molecular biologist can extract just a simple molecule of the DNA from any specimen, an unlimited number of copies will shortly be available for sequencing, comparison with other sequences, and functional annotation (identifying the part of a gene, if any, it codes for). Naturally the older a specimen the smaller amount and the shorter the DNA molecules recoverable. Moreover, in sequencing hominid DNA, the greatest stumbling block is contamination with contemporary human DNA, which literally spews from the fingertips of the investigator running the PCR procedure (Pääbo 1999). But (as yet unreplicated) claims of successful amplification and sequencing include 80-million-year-old dinosaur bones, and 130-million-year-old insects trapped in amber (cf. Hoss 2000). So, if sociality is encoded in the genes, and we can find the right specimens, gene sequencing holds out the prospect of answering questions that are otherwise open only to speculation.

But what reason is there to suppose that either of these two assumptions obtain? What indeed would we be looking for, were we to seek genes for cooperation? The first problem is to characterize the phenotype with sufficient precision. Identifying anatomical phenotypes used to be no trivial matter, and identifying behavioral ones, if they exist, is much trickier. As we shall see, in some cases identifying phenotypes has become much simpler in the advent of "positional gene cloning." The strategy involves identifying a trait – usually a disease or deficiency that appears to assort in accordance with Mendelian principles, locating a chromosomal marker in victims that does so as well, and then by automated means zeroing in on the specific gene sequence whose mutation or rearrangement is associated with the trait. We thus establish the normal sequence as the gene for the normal trait in the "normal environmental range." We may be lucky enough to discover that cooperation or some phenotypic component of it can be identified as a phenotype without question by molecular biological techniques.[3]

Suppose the disposition we seek to identify as a phenotype is something like "the disposition to engage in tit-for-tat (TFT) strategies in iterated prisoners' dilemma (IPD) games," or "the disposition to ask for half in iterated cut-the-cake games," or again, "the disposition to reject anything less than half in an ultimatum game." Call the first of these dispositions "TFTinIPD" for short, and the last "1/2RUG." Now, no one supposes that either of these dispositions is a single gene-controlled phenotype like tongue rolling. Genes just don't seem likely to code for recognition of a complex environmental conditional setup in which an abstractly described strategy is to be employed. Rather, if anything like this conditional disposition is a phenotype selected for, then there is a package of genes for much simpler traits on which the disposition supervenes. Several models for such a package of genes suggest themselves.

The first of these is the supervenience of an apparent TFTinIPD disposition or 1/2RUG disposition on simpler behavioral dispositions nature has had a chance to hard-wire: seek genes for a simpler behavior that, in the circumstances where hominids find themselves, produces the same actual behavior as the complex disposition. Consider the male mouse's disposition to kill all

---

[3] It's worth noting, of course, that the expression "gene for" is both widely misunderstood and misused. As I use it here, it can at least be quasi-operationally defined as the sequence that would be identified as the wild type in a positional cloning exercise. The expression "gene for" must be understood as always relativized to a population and an environment.

mouse pups not its own offspring – a highly adaptive bit of environmentally conditional behavior. How can a male mouse discriminate its pups from another male's offspring by the same or closely (related) female(s). Not by look, smell, or other features that a mouse can detect. Male mice have a genetically hardwired pup-killer disposition. But mice do not live in large colonies, and nature equips the male mouse with a further package of genes that programs the mouse to refrain from killing pups it comes across eighteen to twenty days after its last ejaculation. This period happens to be the gestation time for female mice. So pups the male encounters during this period have a high probability of being its own pups and have a chance to escape before the pup-killer instinct returns (Perrigo et al. 1990). For all the world it looks like male mice show a complicated strategy requiring considerable genealogical knowledge, when in fact the behavior is hardwired, and the gene that produces it is a quick and dirty solution to a hard problem. Similarly, the disposition to play an unconditional kin-altruism strategy in iterated prisoners' dilemma circumstances among kin would be genetically simpler to encode and indistinguishable from the complex conditional strategy of TFTinIPD. The trouble with "quick and dirty" solutions like this is that they are vulnerable when circumstances change. Suppose, for example, mice began to live in larger colonies. The quick and dirty solution may become a lethal maladaptation. Similarly primates playing kin altruism will be undone by the emergence in the environment of a free-rider.

That there is a gene for kin altruism or some quick and dirty substitute for it among the mammals and birds is a pretty safe bet. But if there is a "gene for" kin altruism or even any quick and dirty available substitute for it, there is also some considerable evidence that such a gene either never figured among the genotype of primates, or that if it did, it made no significant contribution to cooperation among them. This is due to the fact that long before our last common ancestor with the chimps (about 5 million years ago), all the primates had ceased to live in groups in which kin altruism would be selected for. Or at least that is what a comparative analysis of our closest primate relatives suggests. The social structure of almost all extant ape groups reflects female (and often also male) disbursal at puberty, high uncertainty of paternity (except for gibbons), and an abundance of weak social ties and a lack of strong ones. Paleontology reveals that the number of ape species underwent a sharp decline about 18 million years ago, while monkey species proliferated. If this was the result of competitive exclusion of apes toward marginal tree-limb niches, it would explain many of their and human anatomical similarities. Unlike humans, chimps and gorillas remained in these restricted niches to the present. Humans and chimps are highly individualistic, mobile across wide areas, self-reliant, and independent. By contrast the monkey species reflect matrifocal social networks that would strongly encourage the selection of kin altruism (Maryanski and Turner 1992). At a minimum the pattern of sociality we and the other primates inherited from our last common ancestor makes it highly probable that cooperation among us is not written in the genes, even imperfectly, approximately, by some quick and dirty exploitation of an already available gene for kin altruism, still less by direct natural selection for the disposition TFTinIPD.

It would be more reasonable to assume that TFTinIPD or 1/2RUG is a behavioral disposition simply reinforced by its environment, that is, a disposition ontogenetically selected for, though not philogenetically selected for; then we seek a package of genes that produce the dispositions and capacities individually (nontrivially) necessary but not jointly sufficient for the TFTinIPD or 1/2RUG behavior. (A gene is nontrivially necessary for a phenotype roughly if it is not also necessary for a large number of other traits, including respiration, metabolism, reproduction, survival, etc.) In this scenario a great deal of the burden of explaining the exact shape of cooperation is shouldered by the environment in which hominids must have survived for millions of years. And the degree to which our genomes are explanatorily relevant to cooperative dispositions will turn on whether the genes that subserve cooperation were selected for owing to the fact that they make hitting on TFTinIPD or 1/2RUG overwhelmingly likely and easy to learn from others, or because they make discovering and learning any complex behavior easy. The latter case undercuts the notion that cooperation is an evolutionary adaptation naturally selected for. The former case, on the contrary, would go some way toward vindicating evolutionary scenarios for cooperation.

Evolutionary psychologists like Cosmides and Tooby endorse the former hypothesis based on their striking experiments with the Watson card problem. Other evolutionary psychologists suggest that emotions such as shame and anger were selected for because those organisms that manifest them will be more inclined to play 1/2RUG even when they recognize that doing so reduces their immediate payoffs. By inducing them to play 1/2RUG and communicating the credibility of the resolve to do so in the future, a gene for emotions like anger will be selected for when 1/2RUG is an adaptive strategy. And, collaterally, the emotion of shame will be selected for if it weakens the disposition to demand or take more than half in an ultimatum game.

Exponents of an evolutionary account of cooperation will favor the second of these alternatives according to which dispositions that specifically subserve cooperation are selected for just because they do so. Indeed, some will hold that dispositions and capacities useful for other purposes beside fostering sociality, like memory, speech, and reasoning, have been selected for owing to their contribution to solving the design problem presented by iterated prisoners' dilemmas, and other competitive games. Suppose the genes for a suite of widely useful capacities such as speech, memory, and a theory of mind were all selected for because together they made an agent's seeing and choosing the cooperative strategy a "no-brainer" move in appropriate circumstances. We might be tempted to say that together the sequences do constitute "a gene for cooperation."

But which of these three possibilities obtains is something gene sequencing too may illuminate. We know that the genes needed for the evolution of cooperation will include those which subserve general capacities such as memory, reasoning, and speech and ones specific to cooperation such as the emotions like anger, shame, resentment, guilt, love, jealousy, and revenge.

THE EXPLANATORY CONTRIBUTION OF HEREDITARY AND GENETIC DEFECTS IN HUMANS

One of the ways to begin to identify the relevant phenotypes and genotypes on which cooperative behavior supervenes is to examine hereditary and genetic defects in humans. For example, it has recently been shown that certain significant defects in speech assort in genetically familiar patterns, and positional cloning has enabled geneticists to locate the particular genes responsible for the defect and, *mutatis mutandis,* the genes whose normal function is necessary for normal speech (Lai et al. 2001). It occurred to the researchers making this discovery almost immediately that sequence comparisons with chimps could reveal important information about the evolution of language competence, a vital necessity for the emergence of complex cooperative dispositions. We know that chimps and gorillas have shown substantial communicative behavior in domestication, and ethological study of vervet monkeys continues to increase our knowledge of their lexicon well beyond the well-known calls for eagle, leopard, and snake. What they appear to lack is syntactic skills, and that these skills are genetically hardwired in us is suggested not just by Chomsky's speculations but by Derrick Bickerton's studies of the transition from pidgins to Creoles. The kind of skill involved is one for which a gene has now been identified, localized, and sequenced and is ready for comparison.

For another example more directly tied to the specific dispositions involved in cooperation, consider high-function autism and Asperger's syndrome. They prevent normal cooperative behavior, are associated with anatomical and neurological abnormalities in the brain, and (in the case of autism at least) have a substantial hereditary component. There is reason to suppose that autism results from the interactive effects of at least three micro-rearrangements on genes, some of which produce a serotonin transporter. These genes are probably located on chromosomes 7 and 15, and they are implicated in some other rare genetically caused retardation.

The possibility should not be neglected that these syndromes involve defects in a capacity already subject to some interspecies investigation. We know that normal children develop a "theory of mind" – the attribution of intentional states to others between the ages of two and four, and there has been some empirical investigation and a good deal of debate about whether the primates show a similar capacity. If the capacity to treat others as having intentional states is one lost in autism, then we are on the way to locating the genes that are either nontrivially necessary or perhaps even sufficient for the capacity in humans. Gene sequence comparisons with the primates may then help answer more definitively the question of whether current evidence demonstrates that apes have a theory of mind; enable us to determine differences and limitations on their having such a theory as a function of the structure of the homologous sequences; and, together with other sequence data, give us clues to the evolution of the gene sequences from our last common ancestors with the primates.

The behavioral and anatomical study of a variety of human syndromes can produce data on heritable component phenotypes that make up the disorder, and correlated gene-locations and sequences, long before we know much about the biosynthetic pathways from the gene sequences to the behavioral syndromes. At this point comparative computational genomics may become relevant to answering questions about the evolution of cooperation. Techniques of "positional cloning" now enable geneticists to locate the precise sequence responsible for a phenotypic disorder (and hence for a normal phenotype) in an almost automated process, once a marker associated with the syndrome is detected in the chromosomes of those who bear it.

Once a "gene for" some defect is found, the "gene for" the nondefective capacity is ipso facto established, and comparative genomics can begin. It is well known that the human and the chimp gene sequences are at least 98.7 percent identical, that sequences for many proteins already identified are indistinguishable between *Homo sapiens* and Pan troglodyte. What is more, we need not await complete and 100 percent accurate sequencing of chimp and human DNA to begin to make or even conclusively to conclude many of our comparative inquiries. First, in both cases only about 2 to 3 percent of the sequences code for functional gene products, and in the second place, we can be confident that the significant and interesting differences between us and chimps – in language, cognition, emotion – turn on differences in regulatory genes that control the timing and order in which structural genes produce their protein products. The same must be true about Neanderthal and other retrievable ancient primate DNA: most of it is junk, and even among the functional DNA what is really important are the regulatory gene sequences. Of course, junk DNA is invaluable for determining lines of descent and approximate ages of specimens, and the importance of only a small fraction of the DNA for many questions is a double-edged sword. It means that, on the one hand, only a little DNA is required to account for great differences among species, and, on the other hand, the likelihood of the important sequences being preserved, as opposed to unimportant ones, is proportionately lower as the important sequences are smaller in size.

The challenge and the promise of gene sequencing as a source for information we need about the evolution of behavioral dispositions are that, on the one hand, we do not yet have a firm characterization for the phenotypic behaviors we seek the genetic basis for, and, on the other hand, given their complexity, the only way to individuate these phenotypic traits and identify their component dispositions is by identifying and manipulating candidate gene sequences. The process must move back and forth between molecular biology and behavioral biology.

So, suppose we identify a human DNA sequence that codes for proteins and enzymes whose absence is sufficient, in most environments for, say, "the defective theory of mind" component of (high-function) autism or, even better, for Asperger's syndrome, in which the antisocial component of autism is evinced without linguistic or intelligence related abnormality. We then seek a sequence in chimp DNA homologous with this sequence. If we find nothing even close in sequence, this by itself would be highly significant: we would have identified a gene for some neurological product absent in the chimp, and presumably important for human sociality. In effect, we would have located a macromutation, a hopeful monster, a most unlikely outcome. More likely we will discover that the sequence is represented in the chimp, but with significant differences in copy number, linkages, introns, and so on. Once we have located the homologous gene sequences in the chimp, it will be essential to discover what proteins these sequences code for, and how they are implicated in chimp behavior.

The homologous chimp gene may have a function altogether unrelated to the function of the "gene for" a theory of mind or some other component of (high-function) autism, or it may control aspects of a behaviorally

192

homologous phenotype. Either way, we may learn a great deal from the comparison. If the gene has no particular obviously homologous behavioral role, then we will be able to infer something important about the role of this gene in the etiology of the human capacity to attribute minds to others: namely, that this gene, which in our last common ancestor had some quite different function, was somehow recruited, as a quick and dirty solution to the design problem of ensuring cooperation. The differences in all these regulatory sequences – their gene-copy number, position, intron numbers, and linkage to other genes – between humans and chimps will give the molecular geneticist clues about the probable course of evolution from our last common ancestor with the chimps to our own and the chimp's current genetic endowment. Such clues can in theory be found by DNA amplification among fossil specimens going back almost one-fifth of the way to that common ancestor, but in practice probably only to the Cro-Magnon-Neanderthal split 600,000 years ago. Furthermore, we probably need to learn a great deal more about chromosomal and genetic rearrangement in general before the narrative of molecular events in the evolution of genes for normal social behavior in humans.

Even if cooperation emerged through natural selection, it may be tempting to suppose that a genetic defect like lacking a theory of mind or (high-function) autism results in genetic defects in a wide variety of genes that code for any one of the components of the biosynthetic pathway that results in normal childhood sociality. That we should expect to find a gene that codes for TFTinIPD or 1/2RUG seems preposterous. It is only slightly less preposterous that we should expect to find a single gene that codes for the component disposition to "attribute a mind to animate bodies in your vicinity." When we discover that in autism the problem is something so "trivial" as a defect in the gene that codes for production of a protein required for blood-platelet serotonin digestion, it becomes tempting to complain that the labeling of the sequence in question as "the gene for" autism seems like a tendentious overdescription. And yet research on other animals has made more surprising discoveries of single genes responsible for very complex conditional behaviors involving apparently voluntary and cognitively nontrivial components.

SIMPLE GENES RESPONSIBLE FOR BEHAVIORS IN ANIMALS

Normal nurturant behavior in mice includes creating nests, cleaning pups, retrieving them to the nest, crouching over them to provide warmth, and offering nourishment. Nurturance in mice reflects a capacity normally acquired by males and females after exposure to similar retrieval, cleaning, warming, and feeding behavior in other mice. When the mouse genome is subjected to a "knockout mutation" of *FosB*, a gene that codes for a 4.5-kilobase messenger RNA, the result is that mothers ignore their pups, do not gather them, retrieve them, warm them, or feed them, although they do approach and sniff them. This behavior remains unchanged through several pregnancies and in the presence of appropriate modeling behavior by wild-type ("normal") maternal mice. So we can exclude learning and experience as causes of infant nurturance. Indeed, wild-type mice that have never been pregnant will show nurturant behavior when exposed to newborn pups, while *FosB* mutant mice who have never been pregnant show the same nonnurturance defect. Nor is the defect even limited to females: wild-type males will nurture, and *FosB* mutant males will not. When subject to tests for cognitive-, olfactory-, or hypothalamic-related abnormalities (hypothalamus defects are known to influence nurturance), the *FosB* mutant mice show no behavioral deficits or abnormalities. Studies of *FosB* gene expression in normal mice brains have led researchers to conclude that exposure to pups triggers the *FosB* gene in cells of the preoptic area of the hypothalamus to produce a protein that appears critical to nurturing. The FosB protein is expressed elsewhere in the brain and may have functions additional to its role in nurturance. However, research has excluded many more basic, and nonspecific roles for *FosB* – in olfaction, general cognition, perception, and learning that might lead to defects in nurturance (and other capacities as well) (cf. J. R. Brown et al. 1996).

It is hard to escape the conclusion that this is a "gene for" nurturing in mice. Why should there not be genes for similarly complex behavior in other mammals, up to and including chimps and humans? Unfortunately, the best way to tell whether there are such genes is simply not open to us. The regulations under which both institutional review boards for human subjects and animal care committees operate make it unlikely that the protocols under which knockout and gene-insertion experiments proceed will ever be approved for humans or chimps. Nevertheless, it is worth considering what such experiments could show. Take, for example, the "grammar gene," as Pinker (2001) calls it, identified by Lai et al., described previously. Many of the affected humans show normal intelligence; "they have trouble identifying basic speech sounds, understanding sentences, judging grammaticality and other language skills" and have a genetic marker at a locus called the SPCH1 segment of chromosome 7, at a specific regulatory gene *FOXP2*, disrupted in their case by a translocation. The translocation results in the substitution of guanine by adenine in the nucleic acid sequence, and arginine by histidine in the gene product. Two forbidden experiments immediately suggest themselves: locate the homolog of *FOXP2* in the chimp (it must be there, because it is already known to be expressed in the developing mouse cerebral cortex), and either insert a normal human *FOXP2* or some portion of it so that the same regulatory product is produced in the chimp. It is well known that the sorts of regimes already employed to test linguistic competence among chimps reveal a lack of grammaticality in their performance that is required for complex schemes of cooperative behavior. Will the gene insertion make a difference either to the individual chimp's language learning capacity or the enhancement of complex communication among chimps? A second experiment, even less permissible, is to locate the homologous gene in chimps and insert it in human infants and to follow development to determine what sorts of linguistic deficits result.

The same sorts of experiments will repeatedly suggest themselves as positional cloning identifies more and more specific DNA sequences implicated in the development and exercise of those human capacities and dispositions we suspect are necessary for TFTinIPD, 1/2RUG, or some such dispositions. Beyond the limitations imposed on experimentation with human and primate subjects, the only limitation to this strategy seems to be the sheer number of genes and gene products that are implicated in these complex dispositions. The gene for nuturance in the mouse is, we like to think, more likely the exception than the rule among mammals. But even if it is common, the number of gene products involved in complex behavior may well be beyond current computational limits. If upward of 60 percent of the coding regions of the genome are devoted to the production of proteins and enzymes expressed in the brain, then even to identify a significant portion of the "genes for" something as complicated as cooperation will be a vast undertaking. But this fact does not detract from the in-principle possibility of employing gene sequencing to illuminate the evolution of cooperation.

HOW TO EXPLAIN THE EVOLUTION OF COMPLEX HUMAN BEHAVIORAL DISPOSITIONS

Long before we have identified all the genes nontrivially necessary for complex cooperative behavior, we will have identified enough of them to construct a number of macromolecular scenarios for how linkages, crossover events, mutation, gene duplications and translocation, and other events were selected for to produce the nucleotide sequences that result in a complex cooperative disposition in us, and the nucleotide sequences that produce the more limited social dispositions of chimp. From these changes and the determination of what their gene products do, we will ultimately be able to infer some of the

195

changes in environmental conditions that selectively forced such relatively rapid genetic changes. That such genetic alterations hold the key to our distinctive capacities and dispositions is reflected in the fact that, despite the tiny quantitative nucleotide difference between us, the chimps, and the gorillas, they are both relatively unsuccessful species, still restricted to a narrow and endangered niche geographically close to the one we started out in, while we bestride the globe.

With the right hominid fossils and a great deal of good fortune, PCR amplifications can help us narrow the genetic scenarios to a small number. We already know that the sequence that the mitochondrial and nuclear DNA sequence differences between Cro-Magnon and Neanderthal are too great for there to have been interbreeding between them. Sequencing for genes whose defects make for noncooperative behavior in us will be singularly illuminating. If they are missing, we may have an explanation for why Neanderthal became extinct and we did not. And with a lot of good luck we can make the same comparison between us and hominids who died out long before our appearance. If all the differences, or the largest differences, turn out to be found in those genes implicated in cooperation, the conclusion will be strengthened that at some point in the evolution of humankind cooperation became a forced move or a neat trick that our ancestors found in design space. Probably nothing would be more substantial a biological explanation and vindication of human morality than such a result.

Even if the fossil record is too gappy, or the custodians of the fossil crania prohibit their being ground up in the search for ancient DNA, or we are just plain unlucky in our search for the relevant sequences, there remains one certain source of the evidence that will test the thesis that cooperation, and the genes that made for it, emerged through natural selection. Assume that working from hereditary human defects, employing positional cloning, we have identified and located the group of genes for linguistic communication, or grammaticality, and the group of genes for a theory of mind, and the group responsible for the emotions crucial to commitment, and the group of genes for memory and reidentification of fellow players in competitive and cooperative games, or simply assume that we have identified many of these groups and some of the genes in each group. Then, as with the mitochondrial DNA sequences, we can compare the particular sequences of large numbers of diverse *Homo sapiens* individuals for the amount of variation in neutral nucleotides within these genes, ones that change over time as a result of point mutation but have no selective consequences. Those groups whose genes show the largest amount of neutral variation will be the oldest in our lineage, and within each group those gene sequences which show the most neutral variation will be the oldest. Moreover, in addition to ordering all these genes and groups chronologically, we should be able to date them against a molecular clock.

Now consider what to infer from various results: suppose that all of these genes and groups of genes are about equally old. Then it is reasonable to believe that they were all selected because they were all involved in solving a design problem, or a small number of connected design problems. If we can correlate the concerted emergence of these genes with what we know about environmental changes, paleoarchaeology, and demography, the conclusion that they were selected for solving one or a small number of connected problems is further strengthened. Suppose, on the other hand, within groups and between groups there is no interesting chronological pattern of appearance of various genes. In that case, it would be reasonable to suppose that they each made an independent but cumulating contribution to the fitness of our lineage. It is pretty clear which of these hypotheses would more strongly support the hypothesis that, if not a forced move, the evolution of cooperation was a neat trick.

REFERENCES

Boyd, R., and Silk, J. 2000. *How Humans Evolved*. New York: Norton.

Brown, J. R., et al. 1996. A Defect in Nurturing in Mice Lacking the Immediate Early Gene *fosB*. *Cell* 86: 297–309.

Brown, T. A., Allaby, R. G., Sallares, R., and Jones, G. 1998. Ancient DNA in Charred Wheats: Taxonomic Identification of Mixed and Single Grains. *Ancient Biomolecules* 2: 185–184.

Cann, R. L. 2001. Genetic Clues to Dispersal in the Human Populations: Retracing the Past from the Present. *Science* 291: 1742–1748.

Dennett, D. 1995. *Darwin's Dangerous Idea*. New York: Simon and Schuster.

Gibbons, A. 2001a. The Riddle of Co-existence. *Science* 291: 1725–1729.

Gibbons, A. 2001b. The Peopling of the Pacific. *Science* 291: 1735–1737.

Gould, S., and Lewontin, R. 1979. The Spandrels of San Marco and the Panglossian Paradigm – a Critique of the Adaptationist Programme. *Proceedings of the Royal Society of London B* 205: 581–598.

Hedges, B. 2000. A Start for Population Genomics. *Nature* 408: 652–653.

Hoss, M. 2000. Ancient DNA: Neanderthal Population Genetics. *Nature* 404: 453–454.

Ke, Y. et al. 2001. African Origin of Modern Humans in East Asia: A Tale of 12,000 Y Chromosomes. *Science* 292: 1151–1153.

Lai, C. S. L., Fisher, S. E., Hurst, J. A., Vargha-Khadem, F., and Monaco, A. P. 2001. A Forkhead-Domain Gene Is Mutated in a Severe Speech and Language Disorder. *Nature* 413: 519–523.

Maryanski, A., and Turner, J. 1992. *The Social Cage: Human Nature and the Evolution of Society*. Stanford: Stanford University Press.

Pääbo, S. 1999. Human Evolution. *TCB* 9: m13–m16.

Perrigo, G., et al. 1990. A Unique Timing System Prevents Male Mice from Harming Their Own Off-Spring. *Animal Behavior* 39: 535–539.

Pinker, S. 2001. Talk of Genes and Vice Versa. *Nature* 419: 465–466.

Renfrew, C., Foster, P., and Hurles, M., 2000. The Past within Us. *Nature Genetics* 36: 253–254.

Richards, M., et al. 2000. Tracing European Flounder Lineages in the near Eastern mtDNA Pool. *American Journal of Human Genetics* 67: 1251–1276.

Sahlins, M. 1974. *The Uses and Abuses of Biology.* Ann Arbor: University of Michigan Press.

Sober, E., and Wilson, D. S. 1998. *Unto Others: The Evolution of Altruism.* Cambridge, Mass.: Harvard University Press.

Stoneking, M., and Soodyall, H. 1996. Human Evolution and the Mitochondrial Gene. *Current Opinion in Genomics and Development* 6: 731–736.

Stumpf, M., and Goldstein, D. 2001. Genealogical and Evolutionary Inference with the Human Y-Chromosome. *Science* 291: 1738–1742.

Turner, C. L., Grant, A., Bailey, J. E., Dover, G. A., and Barker, G. W. W. 1998. Patterns of Genetic Diversity in Extant and Extinct Cattle Populations: Evidence from Sequence Analysis of Mitochondrial Coding Regions. *Ancient Biomolecules* 2: 235–250.

Wilson, E. O. 1975. *Sociobiology: The New Synthesis.* Cambridge, Mass.: Harvard University Press.

Index

Index